What's
Happening
in the
Mathematical
Sciences

Volume 5

What's Happening in the Mathematical Sciences

BARRY CIPRA

EDITED BY PAUL ZORN

AMERICAN MATHEMATICAL SOCIETY

2000 *Mathematics Subject Classification*:
00A06

ISBN 0-8218-2904-1
ISSN 1065-9358

This publication was created using the QuarkXPress® desktop publishing system on a Macintosh G4 computer. Halftones were created from original photographs with Adobe Photoshop, and illustrations were redrawn using Adobe Illustrator on Power Macintosh computers.

Printed at The Herald Press,
Pawtucket, RI.

About the Author: Barry Cipra, who also did the writing for volumes 1–4 of *What's Happening in the Mathematical Sciences*, is a freelance mathematics writer based in Northfield, Minnesota. He is currently a Contributing Correspondent for *Science* magazine and also writes regularly for *SIAM News*, the newsletter of the Society for Industrial and Applied Mathematics. He received the 1991 Merten M. Hasse Prize from the Mathematical Association of America for an expository article on the Ising model, published in the December 1987 issue of the *American Mathematical Monthly*. His book, *Misteaks...and how to find them before the teacher does...* (a calculus supplement), is published by AK Peters, Ltd.

About the Editor: Paul Zorn, who also did the editing for volumes 2–4 of *What's Happening in the Mathematical Sciences,* is Professor of Mathematics at St. Olaf College in Northfield, Minnesota. He received the 1987 Carl B. Allendoerfer Award from the Mathematical Association of America for an expository article on the Bieberbach conjecture. He is former editor of *Mathematics Magazine*.

Cover: A plot of the inner solar system showing the main asteroid belt (green dots) between the orbits of Mars and Jupiter. Celestial mechanics studies the mathematics of gravitationally attracting bodies. The precise dynamics of N bodies is unknown, in general, for N greater than 2, but researchers have recently discovered a slew of new orbital possibilities. (Figure courtesy of the Minor Planet Center at the Smithsonian Astrophysical Observatory. The plot, shown here for November 27, 2001, is updated daily at MPC's website, http://cfa-www.harvard.edu/iau/lists/InnerPlot.html.)

Contents

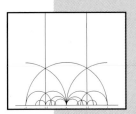

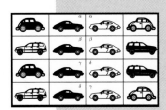

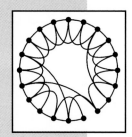
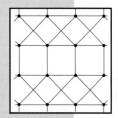
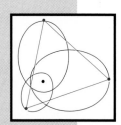

Introduction

Mathematicians like to point out that mathematics is universal. In spite of this, most people continue to view mathematics as either mundane (balancing a checkbook) or mysterious (cryptography). This fifth volume of the *What's Happening* series contradicts that view by showing that mathematics is indeed found everywhere—in science, art, history, and our everyday lives.

Mathematics enables science. The human genome was sequenced using mathematics; the next great task is to understand the products of genes—the proteins—and understanding their structure will require new mathematics (page 12). Centuries ago, mathematicians wrote down the equations that govern the motions of the planets, yet we continue to learn new information about the complicated dance of heavenly bodies (page 68). And cosmologists have spent the past 50 years sketching the history of the universe, but only now do we have the mathematical tools to understand the shape of the universe (page 32).

Mathematics affects our everyday lives. Whether sitting in a traffic jam (page 42) or packing spheres (page 22) or analyzing our collection of acquaintances (page 60), mathematics helps us to understand (and sometimes to improve) the world around us.

Mathematics is poetry. The recent proof of the centuries-old Fermat problem by Andrew Wiles has led to even more spectacular breakthroughs (page 2). The number theory is pure poetry—elegant and engaging. Four thousand years before, an unknown mathematician inscribed other elegant number theory on a clay tablet, and we only now understand the extent of the accomplishment (page 54). One wonders how mathematicians in 4000 years will decipher and appreciate the work of Wiles.

Mathematics is excitement. Thousands of mathematicians engage in that excitement every day. Some of their research will produce better eye glasses or better airplanes next year, but much research will create the foundations for applications fifty or even one hundred years in the future. To emphasize the vitality of mathematics, the Clay Institute offered million-dollar prizes for each of seven outstanding problems (page 76)—just to remind us that vastly more mathematics is yet to be done.

Science, traffic patterns, poetry, excitement. Mathematics is all these things and more. For most people, however, the breadth and depth of mathematics remains inaccessible and the connections above are only dim shadows. *What's Happening* tries to illuminate those shadows for anyone who is curious.

John Ewing
Publisher, American Mathematical Society

1

Modular Cabinet. *Richard Pink, a mathematician at the Swiss Federal Institute of Technology (ETH) in Zurich, Switzerland, designed his own cabinet based on the modular group of 2×2 matrices (see Figure 2, page 4). The cabinets were built by German craftsman Urban Brenner. (Photo courtesy of Richard Pink.)*

New Heights for Number Theory

Andrew Wiles's 1994 proof of Fermat's Last Theorem was, without a doubt, a high point of modern mathematics, an achievement comparable to Sir Edmund Hillary's 1953 ascent of Mount Everest. But the Everest analogy falters in one important respect. For mathematicians, there is no Everest: Each peak is surpassed by another. Challenging as it was, Fermat's Last Theorem can be viewed as a foothill, not a summit, of mathematics.

Number theorists have already begun to reach some of that higher ground. Building on techniques that Wiles introduced in proving Fermat's Last Theorem, researchers have completely solved the Taniyama–Shimura conjecture, one of the central problems in the theory of elliptic curves. They have also made progress on the Langlands Program, a scheme of conjectures that tie together numerous deep results in algebra, analysis, and number theory.

Originally posed in 1955 by Japanese mathematician Yutaka Taniyama, the Taniyama–Shimura conjecture is a bold assertion about two seemingly disparate number-theoretic objects: modular forms and elliptic curves. An elliptic curve can be defined as the solution set of a cubic equation in two variables with rational coefficients, such as $x^3 + y^3 = 1$. Of particular interest are the *rational* solutions of the equation, if any (see Figure 1). The theory of elliptic curves has, in effect, grown up around the problem of studying such sets of rational solutions.

A modular form, on the other hand, is an analytic function, defined

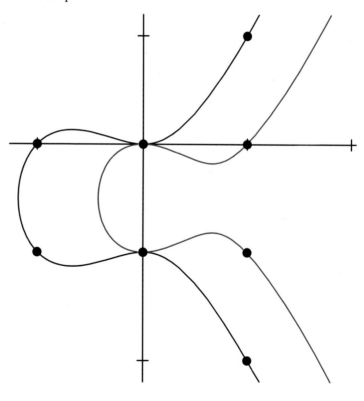

Figure 1. *In the theory of elliptic curves, a single sign can make a world of difference. The curve $y^2 + y = x^3 - x^2$ (teal) has only 4 rational solutions. The curve $y^2 + y = x^3 + x^2$ (black) has infinitely many, only a few of which (with integer coordinates) are shown.*

on the upper half-plane of complex numbers (see Figure 2) that exhibits certain symmetries. One of the symmetries is simple translational invariance: $f(z+1) = f(z)$. (This symmetry appears also in the familiar trigonometric function $\sin 2\pi\theta$.) Others, such as $(11z+1)^{-2} f(z/(11z+1)) = f(z)$, are more recondite, the sort of symmetry it takes years of graduate work to appreciate.

Each elliptic curve and each modular form has an associated complex-valued function known as an L-function. For an elliptic curve, the L-function encodes information about the number of solutions of the cubic equation over finite fields (see Figure 3). For modular forms, L-functions re-express the symmetries in a way that highlights their number-theoretic significance. In both cases, the L-functions have properties similar to those of the famous Riemann zeta function (see "Think and Grow Rich," page 76). In particular, the L-functions can be written as infinite products, with one factor for each prime number.

This form of an L-function is called an Euler product, named for

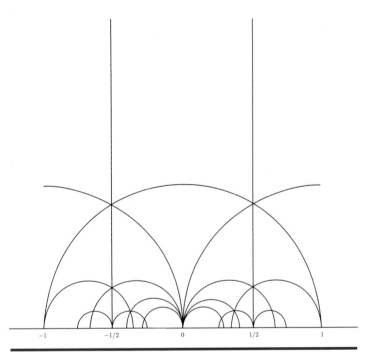

Figure 2. *The upper half plane of complex numbers is tessellated into non-euclidean triangles by the modular group of 2×2 matrices. Every point $z = x + iy$ in the upper half plane can be sent to each triangle by a map of the form $(az+b)/(cz+d)$, where a, b, c, and d are integers satisfying $ad - bc = 1$.*

the eighteenth-century mathematician Leonhard Euler, who noted the corresponding formula for the zeta function. The L-functions for elliptic curves are *defined* as Euler products. Actually, not every modular form has an L-function with an Euler product. To be more precise, modular forms belong to finite-dimensional vector spaces, each of which has a distinguished basis consisting of eigenvectors for a special set of linear transformations known as Hecke operators, named for the number theorist Erich Hecke, who introduced them around 1936. It's the L-functions for these "Hecke eigenforms" that have Euler products. (This paragraph is rated LA, for Linear Algebra. Readers who have not had a course in this subject may wish they had skipped over it.)

In the early 1950s, Martin Eichler of the University of Basel proved that the elliptic curve $y^2 + y = x^3 - x^2$ and a certain modular form have exactly the same L-function—a major discovery at the time. Taniyama's daring conjecture was that this was no accident: The L-function of *every* elliptic curve is also the L-function of a Hecke eigenform.

First stated in a somewhat vague way, Taniyama's conjecture was sharpened by his friend and colleague Goro Shimura, now at Princeton University. Shimura added the stipulation that the eigenforms in question are ones whose symmetries are associated with the so-called "congruence" subgroups of the "modular" group. The

> **The L-function of *every* elliptic curve is also the L-function of a Hecke eigenform.**

p	2	3	5	7	11	13	17
N_p	4	4	4	9	–	9	19
a_p	-2	-1	1	-2	–	4	-2

$$L_E(s) = \frac{1}{(1 - 11^s)} \prod_{p \neq 11} \frac{1}{(1 - a_p p^s + p^{1-2s})}$$

$$= 1 - \frac{2}{2^s} - \frac{3}{3^s} + \frac{2}{4^s} + \frac{1}{5^s} + \frac{2}{6^s} - \frac{2}{7^s}$$

$$- \frac{2}{9^s} - \frac{2}{10^s} + \frac{1}{11^s} - \frac{2}{12^s} + \frac{4}{13^s} + \cdots$$

Figure 3. *The L-function of an elliptic curve is defined by an Euler product, taken over primes not dividing the conductor, which for $y^2 + y = x^3 - x^2$ is 11 (bottom). In the table (top), N_p counts the number of solutions when the equation is taken $\mathrm{mod}\, p$, and $a_p = p - N_p$. For example, when $p = 7$, there are 9 solutions—$(0,0)$, $(0,6)$, $(1,5)$, $(2,4)$, $(3,6)$, $(4,4)$, $(5,5)$, $(6,0)$, and $(6,1)$—so $N_p = 9$ and $a_p = 7 - 9 = -2$.*

modular group, usually denoted Γ, consists of 2×2 integer matrices, $\left(\begin{smallmatrix} a & b \\ c & d \end{smallmatrix}\right)$, with $ad - bc = 1$. For each positive integer N, the congruence subgroup $\Gamma_0(N)$ is the set of matrices in Γ with c divisible by N.

In the lingo of modular forms, the number N is called the "level" of the congruence subgroup. There is a similar notion for elliptic curves, except that the governing number is called the "conductor" of the curve. For $y^2 + y = x^3 - x^2$, the conductor is 11. For $x^3 + y^3 = 1$, it's 27.

In a paper published in 1967, Andre Weil of the Institute for Advanced Study in Princeton further refined the Taniyama–Shimura conjecture, proposing that elliptic curves of conductor N correspond, via their L-functions, to eigenforms of level N.

Weil also found persuasive evidence for the conjecture's validity. (Amusingly, he left the conjecture itself as "an exercise for the interested reader.") Further evidence was given by Shimura. In a paper published in 1971, he proved that every eigenform has an

Richard Taylor. (*Photograph from* The Fermat Diary, *by C.J. Mozzochi, published by the American Mathematical Society, 2000. Provided by the author and reprinted with permission.*)

elliptic curve with the same L-function. But the other half of the conjecture remained: Are there any elliptic curves whose L-functions are *not* L-functions of modular forms?

By the 1970s, the Taniyama–Shimura conjecture was recognized as one of the keys to the theory of elliptic curves. It plays such a central role because the arithmetic secrets of elliptic curves are thought—and in some cases known—to be revealed in the analytic properties of the curves' L-functions, and those properties are most easily examined if the L-functions come from modular forms. Then, in 1985, Gerhard Frey at the University of the Saarland in Saarbrücken, Germany, raised the stakes even higher: The Taniyama–Shimura conjecture, he pointed out, implies Fermat's Last Theorem.

Fermat's Last Theorem, of course, is the famous assertion, posed around 1635, that the equation $x^n + y^n = z^n$ has no solutions in positive integers x, y, and z if the exponent n exceeds 2. For simple reasons it need only be proved if the exponent n is either 4 or an odd prime p. The case $n = 4$ is "easy" to prove;

Christophe Breuil. *(Photograph from* The Fermat Diary, *by C.J. Mozzochi, published by the American Mathematical Society, 2000. Provided by the author and reprinted with permission.)*

Fermat himself gave a proof. Before Wiles's proof, number theorists had managed to prove Fermat's claim for all prime exponents up to around 4 million (see "A Truly Remarkable Proof," *What's Happening in the Mathematical Sciences*, Volume 2). But none of the methods could handle all cases at once.

What Frey observed is that if there were positive integers a, b, and c that satisfy the equation $a^p + b^p = c^p$ (with p an odd prime), then the elliptic curve $y^2 = x(x - a^p)(x + b^p)$ would have some extraordinary properties, among which was that it would *not* correspond to a modular form. In other words, if Fermat's Last Theorem were false, then the Taniyama–Shimura conjecture would also be false. Or, conversely, if the Taniyama–Shimura conjecture is true, then so is Fermat's Last Theorem.

Frey's link between Fermat's Last Theorem and the Taniyama–Shimura conjecture was firmly established by Ken Ribet at the University of California at Berkeley. Ribet's result cemented the relationship between these two daunting problems. But the daunting lasted less than a decade: In 1993, Wiles announced a proof of Fermat's Last Theorem, based on inroads into the Taniyama–Shimura conjecture. His initial proof turned out to contain a mistake, but within a year Wiles, with the help of Richard Taylor at Harvard University, had fixed the flaw.

Wiles's strategy was to prove the Taniyama–Shimura conjecture not for *all* elliptic curves, but only for a subclass known as the "semi-stable" curves. These are elliptic curves whose conductors are square-free (i.e., not divisible by the square of any prime, such as 4, 9, 25, etc.). This is enough to prove Fermat's Last Theorem, because one of the properties of Frey's curve is that, should there be such a curve, its conductor would be the largest square-free divisor of the product abc, and so the curve would be semi-stable.

The long-sought proof of Fermat's Last Theorem electrified the mathematical world. But number theorists

Fred Diamond. *(Photograph from* The Fermat Diary, *by C.J. Mozzochi, published by the American Mathematical Society, 2000. Provided by the author and reprinted with permission.)*

were as much, or even more, ener-
gized by the new ideas Wiles had
used in his proof. The years since
Wiles's proof have seen a surge of
new results in the theory of elliptic
curves. Chief among these is a proof
of the Taniyama–Shimura conjecture
for all elliptic curves.

Announced in 1999, the proof of
Taniyama–Shimura is a joint work of
Christophe Breuil at the Université
Paris-Sud, Brian Conrad at Harvard
University, Fred Diamond at
Brandeis University, and Taylor.
Their paper, "On the modularity of
elliptic curves over **Q**: Wild 3-adic
exercises," appeared in 2001, in
the *Journal of the American
Mathematical Society*.

Brian Conrad. *(Photograph from* The Fermat Diary, *by C.J. Mozzochi,
published by the American Mathematical Society, 2000. Provided by the
author and reprinted with permission.)*

As in Wiles's work, the new
proof's action is centered on properties of "Galois representations."
These are abstract, group-theoretic objects associated with the set
of algebraic numbers—that is, the set of all roots of all polynomi-
als $P(x)$ with rational coefficients. There is a notion of L-function
and conductor for Galois representations that leads to a generalization of the Taniyama–Shimura conjecture. In effect, Taylor and
crew proved this general conjecture for a class of Galois representations large enough to encompass all elliptic curves.

But the relationship of Galois representations and modular
forms runs even deeper. The full extent of the relationship and its
implications constitutes what's called the Langlands Program,
named after Robert Langlands of the Institute for Advanced Study
in Princeton, who laid out a series of conjectures in the 1960s.
Roughly speaking, the Langlands Program generalizes the notion
of modular forms to functions called automorphic forms, whose
symmetries are associated with $n \times n$ matrices, and it considers
Galois representations with the same class of matrices. The conjectures are highly technical, but basically they assert the existence
of a one-to-one correspondence between modular forms and Galois
representations for each value of n.

The Taniyama–Shimura conjecture fits within the Langlands

In many theories that make claims for the positive integers, the first case or two is either trivial or runs counter to the rest of the theory.

Program for $n = 2$. In many theories that make claims for the positive integers, the first case or two is either trivial or runs counter to the rest of the theory. (In Fermat's Last Theorem, for example, the equations $x + y = z$ and $x^2 + y^2 = z^2$ have infinitely many solutions!) But with the Langlands Program, even the case $n = 1$ is non-trivial; in effect, it entails all of classical algebraic number theory. Its only "trivial" aspect is that a 1×1 "matrix" is simply a number, and numbers, unlike matrices, commute when you multiply them (see Figure 4). Roughly speaking, the Langlands Program is a vast non-commutative generalization of a deep but relatively simple commutative theory.

$$2 \times 5 = 10 = 5 \times 2$$

$$\begin{pmatrix} 2 & 3 \\ 3 & 5 \end{pmatrix} \times \begin{pmatrix} 2 & 1 \\ 3 & 2 \end{pmatrix} = \begin{pmatrix} 13 & 8 \\ 21 & 13 \end{pmatrix}$$

$$\begin{pmatrix} 2 & 1 \\ 3 & 2 \end{pmatrix} \times \begin{pmatrix} 2 & 3 \\ 3 & 5 \end{pmatrix} = \begin{pmatrix} 7 & 11 \\ 12 & 19 \end{pmatrix}$$

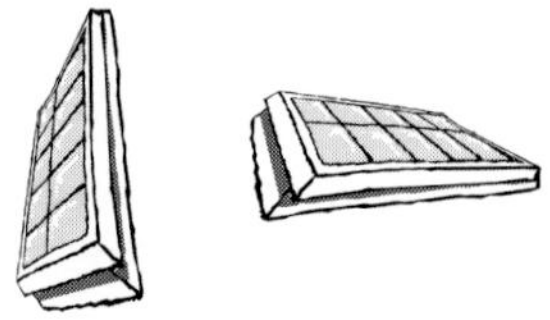

Figure 4. *Numbers commute under multiplication—that is, an ice cube tray holds the same amount of ice no matter how you stick it in the freezer. But matrices, in general, do not.*

Bits and pieces of the program were in place when Langlands formulated his conjectures, and more followed. In addition to its stratification by n, the Langlands Program separates into four main cases according to two properties of the number field involved. A field is simply an algebraic system in which numbers can be added, subtracted, multiplied, and divided (with the exception of division by 0). The most familiar field, the field of rational numbers, is an example of a "global" field "of characteristic 0." Number theory also deals with "local" fields and with fields of positive characteristic. By the mid 1980s, three of the four categories had been settled when $n = 2$, and researchers had made beginning steps for $n = 3$.

The pace has accelerated in recent years. In 1998, Taylor and Michael Harris at the University of Paris VII and, simultaneously, Guy Henniart at the University of Paris-Sud announced complete proofs of the "local" Langlands conjecture—that is, a proof that applies to all local fields and all values of n. At a book-length 200-plus pages, Harris and Taylor's proof presents a detailed description of the correspondence between automorphic forms and Galois representations. Henniart's proof is much shorter but offers less information about the nature of the correspondence.

More recently, Laurent Lafforgue, also at the University of Paris-Sud, has proved the global conjecture for fields of positive characteristic. Lafforgue's work extends methods introduced by Russian mathematician Vladimir Drinfeld, who had proved the result for $n = 2$ in the 1970s. And the proof of the Taniyama–Shimura conjecture takes a step toward proving the Langlands conjecture for global fields of characteristic 0: It shows that many, if not all, Galois representations in this case for $n = 2$ come from modular forms.

These breakthroughs have given mathematicians new tools for studying number theory and its pervasive relationships with other areas of mathematics. But they leave open what many consider the most important case, namely global fields of characteristic 0—and specifically the field of rational numbers. Even for $n = 2$, despite the Taniyama–Shimura breakthrough, "we're a long way from proving the Langlands conjecture," says Taylor. "The original problem we'd really love to understand is still there."

Robert Langlands. *(Photograph from* The Fermat Diary, *by C.J. Mozzochi, published by the American Mathematical Society, 2000. Provided by the author and reprinted with permission.)*

Laurent Lafforgue. *(Photograph courtesy of C.J. Mozzochi and reprinted with permission.)*

"Pax Mundi." *Self-taught sculptor Brent Collins works with geometric ideas—in this case a highly symmetric ribbon surrounding an imaginary sphere. The result is reminiscent of biologists' sketches of proteins. (Photo courtesy of Brent Collins and Phillip Geller Photography.)*

A Mathematical Twist to Protein Folding

Life is basically simple. An acid known as DNA spells out long lists of other acids, known as amino acids. The amino acids link up in their DNA-determined order to form big molecules called proteins. Proteins do many things; the most important is to help turn one DNA molecule into two identical—or nearly identical—DNA molecules.

What's complicated are the details.

There are macromolecular mysteries at every turn. To solve them, molecular biologists are using mathematical and computational methods. And a growing number of mathematicians and computer scientists are turning to problems in molecular biology.

"It's a very attractive area," says Bonnie Berger, a mathematician at the Massachusetts Institute of Technology. Berger wrote her doctoral dissertation on randomized algorithms, which arise in theoretical computer science (see "Random Algorithms Leave Little to Chance," *What's Happening in the Mathematical Sciences*, Volume 2). She now works mainly on problems involving proteins.

One pressing problem in protein chemistry is to determine proteins' physical, three-dimensional structure from the sequences of amino acid "residues" that define them. Biochemists know that each protein "folds" into a specific shape, and that this shape plays a key role in how the protein functions. Knowing a protein's shape enables researchers to better understand the chemical reactions it participates in and, for example, design drugs that target it and it alone.

Protein structure can sometimes be determined experimentally, by X-ray crystallography or nuclear magnetic resonance. But the experimental approach is complicated and time-consuming. It can take a year for a single protein—if it can be done at all. (Many important proteins, such as those that span cell membranes, are only grudgingly amenable to current techniques.) So far researchers have managed to solve the structure of around fifteen

Bonnie Berger. *(Photograph courtesy of Sylvia Allen.)*

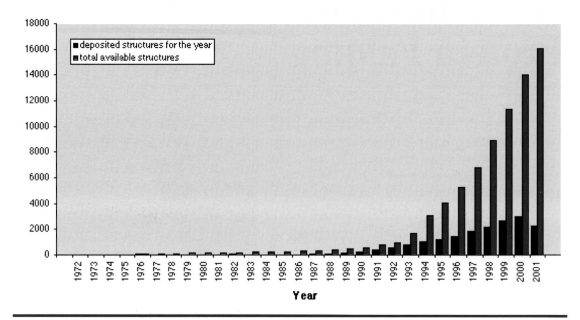

Figure 1. *The number of solved structures in the Protein Data Bank grows annually. (Figure courtesy of the Protein Data Bank, www.pdb.org, updated from "The Protein Data Bank," Berman et al.,* Nucleic Acids Research, *volume 28 (2000), pp. 235–242.)*

thousand proteins (the number grows daily—see Figure 1). These solved structures are stored in the online Protein Data Bank (PDB), operated by the Research Collaboratory for Structural Bioinformatics, a consortium operated by Rutgers University, the San Diego Supercomputer Center at the University of California at San Diego, and the Biotechnology Division of the National Institute of Standards and Technology.

Many of the solved protein structures are simple variants of others, differing only by a single amino acid; experts estimate they understand the structure of only about a thousand different kinds of folds. Considering the immense variety of life, the current protein database is a proverbial drop in the bucket, a bit like having a compendium of the pronouns in Shakespeare's plays.

Computation and mathematical analysis can help on several fronts. In principle it should be possible to predict a protein's shape simply from its sequence of amino acids and the laws of physics. Indeed, researchers in molecular dynamics have had great success setting up equations that describe the interactions that determine the shapes of proteins and other macromolecules. The basic idea is straightforward: A protein adopts the shape that minimizes its overall energy, so solving for its shape is a "simple" problem in multivariable calculus. But the calculus problem may have anywhere from dozens to thousands of variables. And, like the cratered surface of the moon, the energy "landscape" is littered with local minima. The computational problem is to find the *global* minimum among all the local impostors. In fact, a main conundrum of

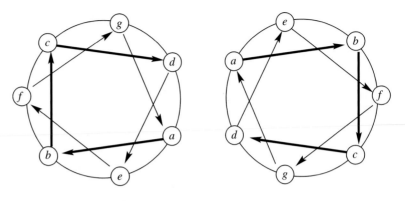

Figure 2. *The geometry of a coiled coil has a natural, 7-residue repeat on each alpha helix.*

protein chemistry is how—if at all—nature manages to solve the problem. Molecular dynamics will probably become an indispensable tool in the study of proteins, but up to now, it has played a limited role.

Berger and her colleagues have taken a different tack. They are banking on the fact that proteins tend to have similar features. Because of this, they say, knowing the structure of one gives information about the structure of others. They call their approach structural motif recognition. The idea is to identify large classes of proteins which are likely to have similar component structures, which can then be verified for proteins of interest. This kind of computational triage helps molecular biologists zero in on what's best to investigate experimentally. "We save them a lot of time in the lab," Berger notes.

The motif recognizers have focused successfully on two structures: coiled coils and beta helices (see Box, "Protein Terminology," next page). They have written computer programs that search efficiently for proteins that include these structures. Their results have revealed, among other things, a previously unknown role for beta helices in bacterial infection.

Coiled coils form when certain "hydrophobic" amino acids are brought into line by the protein's alpha helices. The super-secondary coiling changes the number of residues per turn from 3.6 to 3.5. This produces a 7-unit repetition on each helix. In this arrangement, the lined-up residues appear at positions "a" and "d" (see Figure 2).

Most coiled coils have at least four such 7-unit repetitions. Consequently, one can search a given protein for coiled coils by

The motif recognizers have focused successfully on two structures: coiled coils and beta helices.

Protein Terminology

In trying to understand the 3-dimensional structure of proteins, molecular biologists have developed a special nomenclature. "Primary structure" is simply the sequence of amino acids that make up the protein. "Secondary structure" consists of local folding patterns, which occur in almost all proteins. Chief among these are alpha helices and beta sheets. An alpha helix resembles a telephone cord, with 36 amino acids for every 10 turns of the cord. A beta sheet can be likened to the path you might use in mowing a rectangular lawn. The complete, 3-d shape is called the protein's "tertiary structure."

The hierarchy also includes "super-secondary structures," which are composed from two or more secondary structures. These include "coiled coils" and "beta helices" (see Figure 3). A coiled coil consists of two or more alpha helices twisted about each other, like tangled telephone cords. Coiled coils are common in proteins that bind to DNA and in membrane fusion proteins, which enable viruses such as HIV to enter cells.

A typical beta helix contains three beta sheets forming, roughly, a triangular prism. Instead of weaving each sheet separately, the protein cycles through the sheets, adding one strand at a time to each. Each "rung" of the helix consists of three strands and three turns. There are typically 22 amino acids per rung, making beta helices about six times larger than alpha helices. Thanks to computational biology, there is new evidence that bacteria rely on beta helices to cause infection (see main article).

scanning its sequence of amino acids 28 at a time, comparing them to known coiled coils. From the Protein Data Bank, one can obtain the relative frequency of each residue at each of the seven positions in the known coiled coils. For example, Leucine, which is known (from GenBank, a repository of genetic information) to occur in slightly more than 9% of all positions in proteins, represents around 29% of all position-a's in coiled coils, for a relative frequency of 29/9, or a little over 3. Leucine also has a relative frequency of nearly 4 for position d. By comparison, Proline has *never* been found at position a, even though it accounts for around 5% of all amino acids in GenBank.

The basic idea is to consider a string r_1, r_2,..., r_{28} of residues, and compute a "score" for it—high for coiled coils and low otherwise. Earlier researchers did this simply by taking the product of the relative frequency of r_1 at position a, r_2 at position b, etc., all

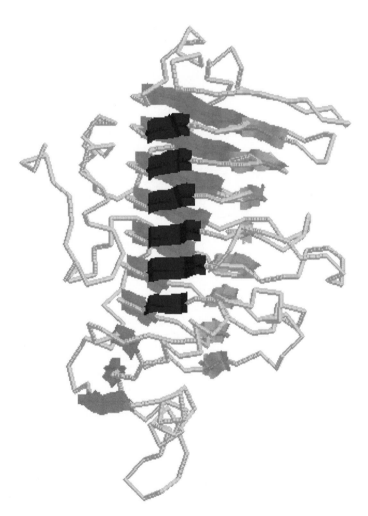

Figure 3. *A coiled coil (top) and beta helix (bottom). (Figure courtesy of Bonnie Berger.)*

the way to r_{28} at position g. For known coiled coils this product is large. But the product also turns out to be large for many strings that are *not* coiled coils; using a large product to indicate coiled coils produces false positives about two-thirds of the time.

In 1995, M.I.T. biologist Peter S. Kim, now Excutive Vice President of Research and Development at Merck Research Laboratories, asked Berger to develop a more discriminating statistical approach. To do so, she used pairwise correlations between amino acids that wind up close to one another if the string forms an alpha helix. Because there are 3.5 residues per turn, residue r_1 interacts with residues r_2, r_4, and r_5. Similarly, r_2 interacts with r_3, r_5, and r_6, and on down the line. (In all, each residue, except those near the two ends, interacts with six others.)

Berger's strategy, implemented in a computer program called PairCoil, is to average the relative frequencies of the three pairs (r_i, r_{i+1}), (r_i, r_{i+3}), and (r_i, r_{i+4}) for each i, and then multiply all 28 averages. (It would be preferable, Berger notes, to use relative frequencies of all four residues at once, but there's not enough data in the Protein Data Bank to permit this. There are $20^4 = 160,000$ quartets of amino acids, but only $20^2 = 400$ pairs.) When fed sequences for the known structures in the PDB, PairCoil separated coiled from non-coiled coils almost perfectly (see Figure 4).

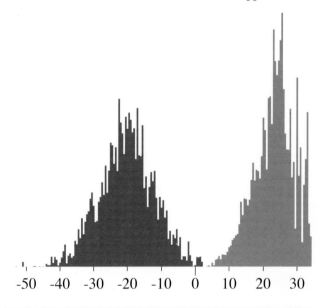

Figure 4. *Histogram of residues scores computed by PairCoil for non-coiled coils in the Protein Data Bank (gray) and 2-stranded coiled coils not in the database (black). (Reprinted with permission from "Predicting coiled coils by use of pairwise residue correlations," Berger et al.,* Proceedings of the National Academy of Sciences, *volume 92, (1995), pp. 8259–8263. © 1995 National Academy of Sciences, U.S.A.)*

Next up were coiled coils consisting of *three* alpha helices. In 1997, Berger and colleagues Kim and Ethan Wolf (who had also collaborated on PairCoil) developed MultiCoil, which assigns *two* scores, measuring the likelihood the string is part of a two- or three-helix coiled coil. The results again show a distinct clustering of scores (see Figure 5).

But PairCoil and MultiCoil weren't perfect. "We found that these programs were missing coiled coils in important proteins such as HIV and other viral membrane fusion proteins, for which there are limited known solved structures," Berger says. The answer came in the form of "learning algorithms," developed in collabora-

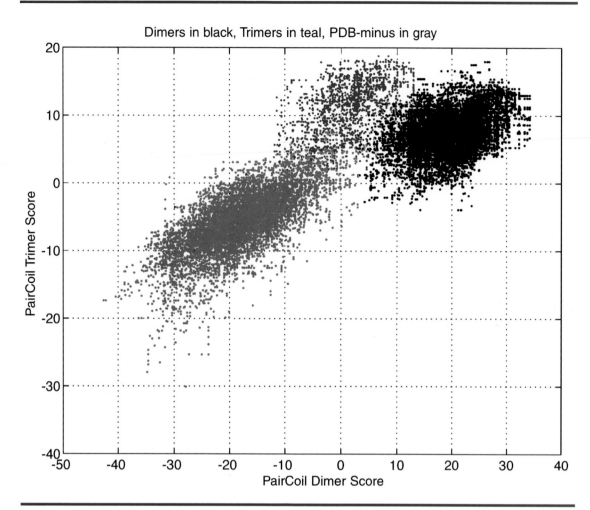

Figure 5. *Scatterplot shows that MultiCoil can distinguish among 2-stranded (black), 3-stranded (teal) and non-coiled coils (gray). (Reprinted with permission from "Multicoil: A program for predicting two- and three-stranded coiled coils," Wolf et al.,* Protein Science, *volume 6 (1997), pp. 1179–1189.)*

tion with Mona Singh, a computer scientist now at Princeton University. They developed two versions of LearnCoil, one for histidine kinase linker domains and one for viral membrane fusion proteins. When run against GenBank sequences for a range of virus families, LearnCoil-VMF found evidence of coiled coils in several, including ebola, influenza, and HIV (see Figure 6, page 21). "Our findings provided further evidence that diverse viruses may in fact share similar methods for viral-cellular membrane fusion," Berger says. "Several of our predictions are now known to be true, including those for ebola virus and human T-cell leukemia virus."

The success of motif recognition techniques for coiled coils has two main sources: the local nature of the interactions in the pattern, and the relative abundance of known examples in the PDB. With beta helices, by contrast, researchers have been less lucky. The PDB includes just 12 examples of this structure, which was first reported only in 1993. And the turns in the rungs have unpredictable lengths, complicating the search for residue pairs that might match up.

Nevertheless, Berger and colleagues Lenore Cowen, now at Tufts University, and Phil Bradley have found an algorithmic approach that works even with the scant data at hand. The key is a short turn, labeled T2, between two of the beta strands that make up each rung of the helix. The T2 turn is usually only two residues long, and its residues have special correlations. The new program, BetaWrap, first identifies a likely T2 turn. Then it estimates the likelihood that the regions on each side will form beta strands. Next, it looks further down the sequence for other strands that have a high probability of forming beta sheets with the first two, and are also separated by a likely T2 turn. Keeping track of several candidates, BetaWrap iterates the process out to five rungs, after which it scores the candidates and assigns the protein the average of the top 10 scores.

Tested on the Protein Data Bank, BetaWrap found all 12 known beta helices, with no false positives—an impressive performance for any algorithm. Sifting through the protein database at the National Center for Biotechnology Information, a division of the National Library of Medicine at the National Institutes of Health, BetaWrap found 2448 likely beta helices (out of 595,890 candidates). There are almost certainly some false positives among them, Berger notes, but the list gives molecular biologists a good starting point.

BetaWrap's results paid an unexpected dividend: Among the 200 top-scoring beta helix candidates, a startling number come from proteins found in human pathogens, including bacteria responsible for cholera, ulcers, malaria, and anthrax (see Figure 7). This suggests that beta helices are important to the mechanisms by which bacteria infect other cells. M.I.T. biologist Jonathan King, who worked with Berger's group, proposes that bacteria use the beta helix to attach to or penetrate the cell membranes of their victims. If molecular biologists confirm BetaWrap's suspicions, the results could spur medical advances in the treatment of countless diseases. Once the structure of these bacterial proteins is understood, Berger says, "maybe we can get some handle on how to stop them."

Arenaviridae	Lassa
Baculoviridae	
Bunyaviridae	Hantavirus
Coronaviridae	
Filoviridae	Ebola
Flaviviridae	Yellow fever
Herpesviridae	Human herpes
Orthomyxoviridae	Influenza
Paramyxoviridae	Mumps, Measles
Retroviridae	HIV
Rhabdoviridae	Rabies
Togaviridae	Rubella

Figure 6. *Viral families examined by LearnCoil, with families predicted to have coiled coils in boldface. Several of the predictions have been verified. Scientific family names (left) and sample family members (right).*

Vibrio cholerae	Cholera
Helicobacter pylori	Ulcers
Plasmodium falciparum	Malaria
Chlamydia trachomatis	Venereal infection
Chlamydophilia pneumoniae	Respiratory infection
Listeria monocytogenes	Listeriosis
Trypanosoma brucei	Sleeping sickness
Borrelia burgdorferi	Lyme disease
Leishmania donovani	Leishmaniasis
Bordetella bronchiseptica	Respiratory infection
Trypanosoma cruzi	Sleeping sickness
Bordetella parapertussis	Whooping cough
Bacillus anthracis	Anthrax
Rickettsia rickettsii	Rocky Mountain spotted fever
Rickettsia japonica	Oriental spotted fever
Neisseria meningitidis	Meningitis
Legionella pneumophila	Legionnaires' disease

Figure 7. *Some human pathogens that BetaWrap predicts to have beta helices. Scientific names (left) and associated human diseases (right).*

Johannes Kepler. *(Courtesy of the Johannes Kepler University of Linz.)*

Nothing to Sphere But Sphere Itself

The mathematician–astronomer Johannes Kepler is best known for his laws of planetary motion, most notably the fact that orbits typically are elliptical. Kepler's revolutionary conclusions stemmed in part from his trust in one of the best computers of his day: Tycho Brahe. Brahe's detailed and highly accurate observations of Mars, made over two decades, led Kepler to reject the circle in favor of the ellipse as the only curve that fit the data.

Mathematicians also remember Kepler for discovering another of nature's laws. This one led to a mathematical problem, known as Kepler's conjecture, which resisted attempts to solve it for nearly 400 years. But the problem seems finally to have been settled. In 1998, mathematician Tom Hales at the University of Michigan announced the completion of a detailed analysis which proves Kepler's conjecture. Fittingly, Hales's solution relies in part on detailed, highly accurate calculations of a modern-day, electronic computer, which ruled out a multitude of alternatives.

The Kepler conjecture concerns the best way to pack spheres together. If you've ever bought oranges at a grocery store or pondered a stack of cannonballs at an old battlefield monument, you've seen the answer: It's achieved by an arrangement that crystallographers call the face-centered cubic packing (see Figure 1). Kepler took note of this in 1611, in "The Six-Cornered Snowflake," an essay on geometric forms in nature.

It's relatively easy to convince yourself that you can't do better than the face-centered cubic packing. But *proving* the conjecture is a whole lot harder—hard enough to warrant its inclusion among the famous "Hilbert Problems," a series of mathematical challenges outlined by David Hilbert in 1900, in a lecture at the second International Congress of Mathematicians.

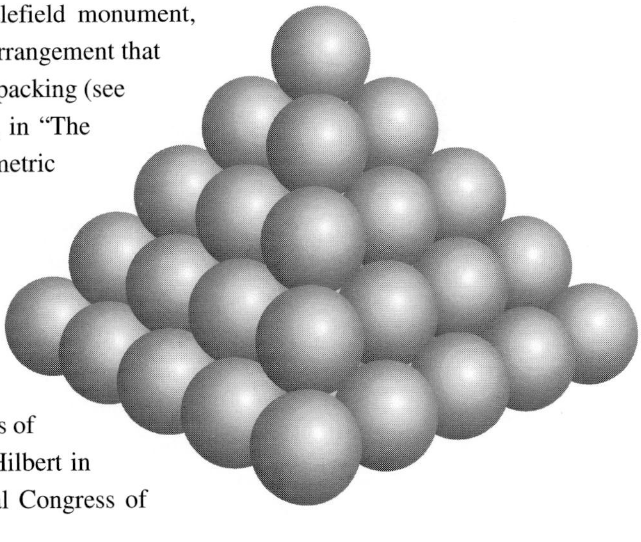

What exactly is the Kepler conjecture? Imagine you have an empty box and a bunch of golf balls. Let's say the box is 20 inches on a

Figure 1. *Face-centered cubic packing. (Figure courtesy of Sam Ferguson.)*

If you start pushing pennies around, it quickly becomes obvious that the best way to fill the plane is with a hexagonal array.

Figure 2. *Hexagonal packing of pennies.*

side and the golf balls are 2 inches in diameter. (This is actually a little big for a golf ball. The official, USGA diameter is 1.68 inches.) How many golf balls can you get in the box and still close the lid?

A simple calculation gives the volume of the box: $20 \times 20 \times 20 = 8000$ cubic inches. A sphere of radius r has volume $V = \frac{4}{3}\pi r^3$, so each golf ball takes up $\frac{4}{3}\pi \approx 4.18879$ cubic inches. So arithmetic says you can't possibly pack more than $8000/\frac{4}{3}\pi \approx 1909.86$ golf balls into that box. (For real golf balls, with $r = .84$, the calculation gives 3222.28.)

But that would happen only if the golf balls occupied *all* the volume of the box, and it's clear that there must be empty space—air—around the balls. The estimate $8000/\frac{4}{3}\pi$ is an upper bound on the answer. So how many balls *can* you cram in the box? What *fraction* of that upper bound is actually possible?

More precisely, the Kepler conjecture asks what fraction of the upper bound is possible as the box gets larger and larger. More precisely still, it asks for the *limit* of the fraction as the sides of the box go to infinity.

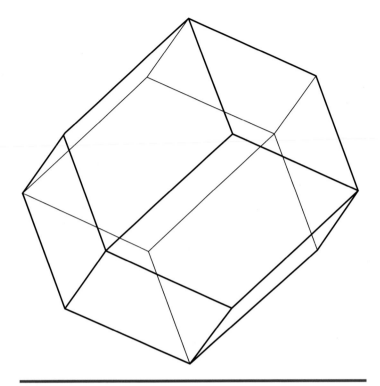

In the 3-dimensional, face-centered cubic packing, each sphere sits in a 12-sided box known as a rhombic dodecahedron.

Figure 3. *Rhombic dodecahedron (also called the Hantzsche–Wendt space—see Figure 2 in "Finite Math," page 36).*

The answer, according to the Kepler conjecture, is $\pi/\sqrt{18}$, or about 74.048049%.

That number is easiest to understand by starting with the two-dimensional analog of the Kepler conjecture: packing circles. If you start pushing pennies around, it quickly becomes obvious that the best way to fill the plane is with a hexagonal array (see Figure 2). Each penny sits inside a regular hexagon. Since the hexagons are all identical, the fraction of the plane occupied by all the pennies is equal to the fraction of a single hexagon occupied by the single penny inside it. But that's easy to compute. If the penny has radius r, then its area is πr^2, while the area of the enclosing hexagon is $\sqrt{12}r^2$. This gives the fraction $\pi/\sqrt{12} \approx 90.69\%$.

In the three-dimensional, face-centered cubic packing, each sphere sits in a 12-sided box known as a rhombic dodecahedron (see Figure 3). For spheres of radius 1, the volume of each box is $4\sqrt{2} \approx 5.65685$, and this gives $\pi/\sqrt{18}$ for the sphere-to-box ratio.

The 2-d hexagon and 3-d rhombic dodecahedron are examples of what geometers call *Voronoi cells*. The points inside a Voronoi cell are, by definition, closer to that cell's sphere (or circle) than to any other sphere (or circle). Every packing determines a system of Voronoi cells (see Figure 4).

One of the first hints of the difficulty of proving Kepler's conjecture is that, whereas in two dimensions the hexagon turns out to be the smallest possible Voronoi cell that can ever occur in any packing, the same is not true for the rhombic dodecahedron. In particular, there are sphere packings in which at least one of the Voronoi cells is a *regular* dodecahedron (see Box, "Hard Work Brings Minimal Result," page 29). The volume of this shape is $20\sqrt{(65 - 29\sqrt{5})/2} \approx 5.55029$—ever so slightly smaller than the volume of the rhombic dodecahedron.

If there were a packing in which *every* Voronoi cell were a regular dodecahedron, then that packing would be a counterexample to Kepler's conjecture. But, just as you can't tile the plane with regular pentagons without some overlapping, you can't fill space with non-overlapping regular dodecahedra. On the other hand, there are packings in which infinitely many of the Voronoi cells are regular dodecahedra, and it's far from obvious that the remaining (infinitely many) cells must be any larger.

Indeed, packing proofs are difficult even in two dimensions. The first proof that the regular hexagon's $\pi/\sqrt{12}$ is the best you can do with circles in the plane was given in 1892, by the Norwegian mathematician Axel Thue.

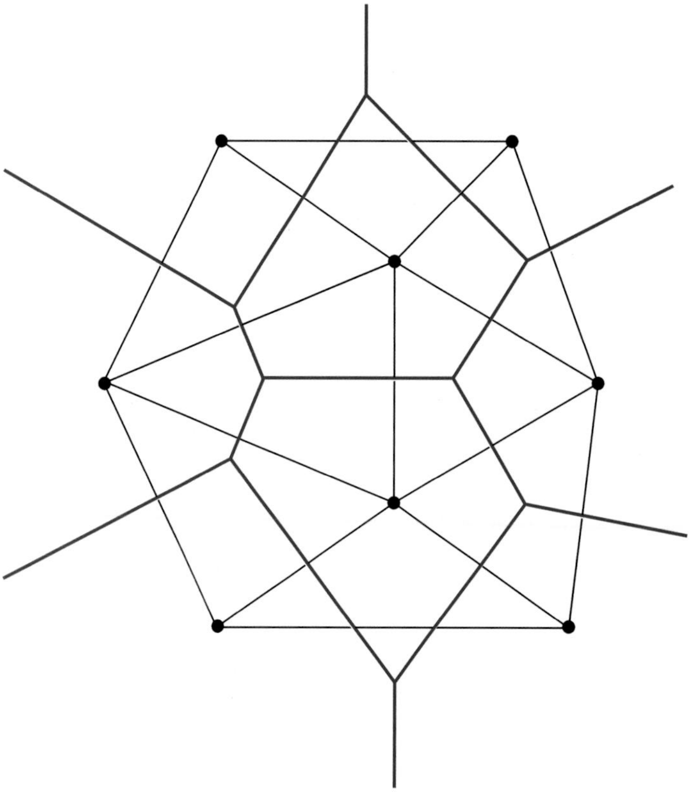

Figure 4. *Voronoi cells for a set of points—representing the centers of 8 circles—in the plane. The sides of the cells (teal) are perpendicular bisectors of lines (black) connecting neighboring points.*

About the only easy thing to prove is that no more than six circles can touch a given circle (see Figure 5). In three dimensions the corresponding number is 12, but this is far from obvious. In fact, it was the subject of a 17th century debate between Isaac Newton, who got it right, and David

Gregory, who thought it might be possible to sneak in an extra, thirteenth sphere. The first rigorous proof was given only in 1953, by K. Schütte and B.L. van der Waerden.

In 1919, Hans Frederik Blichfeldt proved that no sphere packing could take up more than 88.4% of space. A decade later, he lowered this to 83.5%. In 1947, Robert Rankin improved the bound to 82.8%; in 1958, Claude Rogers dropped it further, to 77.97%. This stood for nearly 30 years. Then more reductions came: 77.844% (J. H. Lindsey II, 1986), 77.836% (Doug Muder, 1988), and finally 77.731% (Muder, 1993).

In 1990, Wu-Yi Hsiang at the University of California at Berkeley announced a proof of Kepler's conjecture. Hsiang's proof was published in 1993, but experts in the subject, including Hales, found numerous gaps and flaws in the argument. Hales and Hsiang debated the merits of the proof in the *Mathematical Intelligencer*. The general consensus of the sphere-packing community was that Kepler's conjecture was still open.

So things stood until Hales's recent proof.

Hales, who is now at the University of Pittsburgh, had worked on Kepler's conjecture since the 1980s, but he took on the problem in earnest in 1994. "By late that year I had a pretty detailed strategy of how to go about proving it," he recalls. "Of course it took all the way till '98 to work everything out."

Hales's basic strategy had been outlined in 1953 by the Hungarian mathematician Laszlo Fejes Tóth. Tóth's idea was to reformulate the conjecture in "local" terms, reducing the problem from one about infinitely many spheres to a series of questions about finite arrangements. Tóth even conjectured that computers might play a role in solving the problem, although that was far beyond the vacuum-tube capabilities of the day.

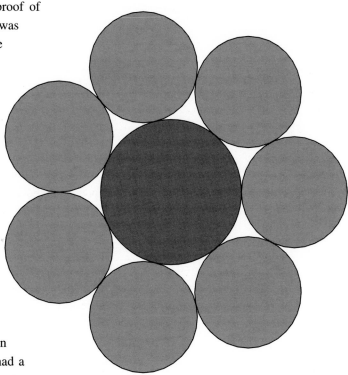

Figure 5. *Who says you can't make seven equal circles touch an eighth?*

Tom Hales. *(Photo by Patricia A. Nagle and courtesy of CIDDE/Photography.)*

Hales began by inventing a new way, which he calls a "star decomposition," to allocate the space around spheres. Roughly speaking, each star is a modified Voronoi cell with some tetrahedral additions. Next, Hales defined a new way to "score" the local density of each star. The face-centered cubic packing turns out to have a score of 8. The definitions are "extremely technical," he says, but their purpose is quite simple: Stars and scores are defined so that any counterexample to Kepler's Conjecture would have to include at least one star with a score above 8.

Then came the *really* hard part: Proving that no star in any packing ever scored more than 8. That's where having a computer came in handy.

"The proof gives a classification of all the decomposition stars that can potentially be a counterexample to the Kepler conjecture," Hales explains. The list is long but finite: A computer program found 5094 different types of stars that might conceivably have a score above 8. Each type needed to be ruled out.

Finding the exact value of the largest score for each type of star, while theoretically possible, was not practical. Fortunately, it wasn't necessary: Hales needed only to show that the scores stayed below 8. To do this, he had the computer convert each inequality estimate into a series of linear programming problems. (Linear programming is a mathematical method widely used for optimization problems in industry.)

To ensure that the numerical calculations didn't let a star with a winning score slip unnoticed through the gaps of round-off error, Hales had the computer do the conversions using a mathematically rigorous technique called interval arithmetic. Roughly speaking, interval arithmetic replaces decimal approximations, such as 0.74048049 for $\pi/\sqrt{18}$, with intervals that contain the numbers being approximated, such as $[0.7404804, 0.7404805]$. Intervals are then added and subtracted by appropriately adding and subtracting their endpoints.

Hard Work Brings Minimal Result

The regular dodecahedron (see Figure 7, page 31) is one of the sticking points for the Kepler conjecture. If all you have is one sphere to house, the regular dodecahedron can do so using less space than its rhombic cousin (see main article). This raises a question of its own: Is there any sphere-housing polyhedron that uses even less space?

The dodecahedral conjecture says no. And this conjecture too is now a theorem. Sean McLaughlin, an undergraduate at the University of Michigan (now in graduate school at New York University), proved it in 1998, using the same computer-intensive approach Hales had taken for the Kepler conjecture.

First formulated by Tóth, the dodecahedral conjecture was expected to be proved as a necessary first step on the way to proving the Kepler conjecture. That's not how things shook out. The approach Hales took sidestepped the issue. But the optimization methods he developed made it clear the dodecahedral conjecture could also be settled.

Not that it was easy. In some ways the Kepler conjecture was easier. "One thing that Sam [Ferguson] and I found very useful was, when we got to a tough point in the proof, we would redesign the optimization problem to avoid the difficulty," Hales says. "You don't have that luxury with the dodecahedral conjecture." So how did McLaughlin deal with the difficulties? Says Hales: "He just had to stay up nights and do it!"

Most of the potential sphere-packing counterexamples were easily disposed of, but some took extra care.

The resulting intervals are guaranteed to contain the corresponding exact values. The trick is to organize the computation so that the intervals remain sufficiently tight.

Most of the potential sphere-packing counterexamples were easily disposed of, but some took extra care. "At the very end it came down to 50 or so cases that the general arguments didn't rule out," Hales recalls. Hales and graduate student Sam Ferguson looked at those cases one by one.

One particularly problematic case, the "pentahedral prism," became the subject of Ferguson's Ph.D. dissertation. Much like the rhombic and regular dodecahedra, the pentahedral prism is associated with a local arrangement of 12 spheres around a thirteenth, central sphere (see Figure 6, next page). But initial calculations only said its central sphere's star had a score less than 10. Hales assigned Ferguson the task of refining the analysis to get a bound below 8. "I only handled one case," Ferguson notes. "It just ended up being the worst case."

Hales's proof is still under review, because of its highly technical nature and its extensive use of computation. At nearly 250 pages, not to mention the gigabytes of computer files, the proof is a daunting read. Hales is currently revising it, mainly to improve the exposition. "I hope that this revision will make things shorter," he says.

The proof's reliance on computers to check the myriad potential counterexamples may someday change. Or maybe not—detailed calculations might prove unavoidable. For now, it's well to recall what Kepler wrote of his own work on planetary orbits: "If you find this work difficult and wearisome to follow, take pity on me, for I have repeated these calculations seventy times...."

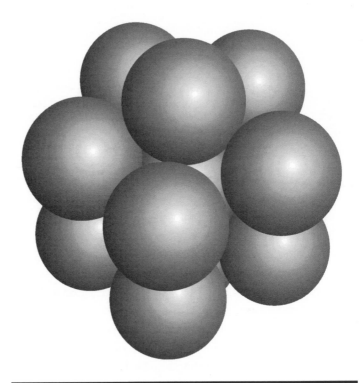

Figure 6. *Sphere packing associated with the pentahedral prism. (Figure courtesy of Sam Ferguson.)*

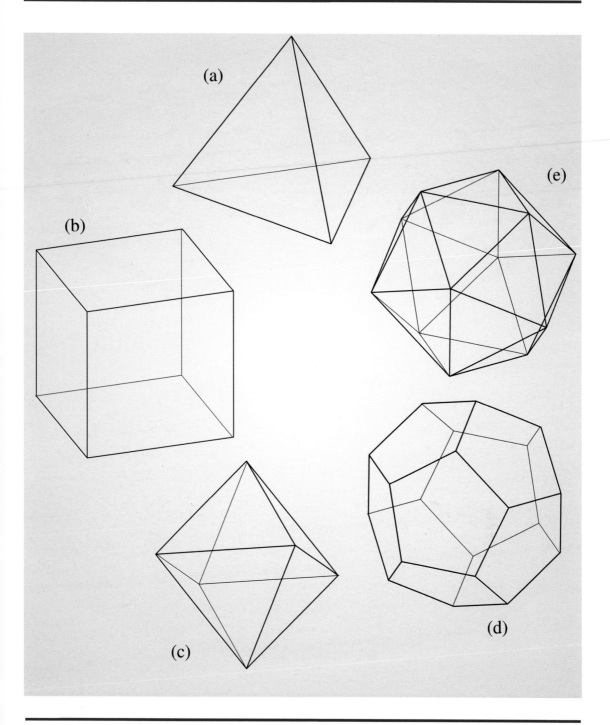

Figure 7. *The regular dodecahedron (d) is one of the five "Platonic solids." The others are the tetrahedron (a), cube (b), octahedron (c), and icosahedron (e).*

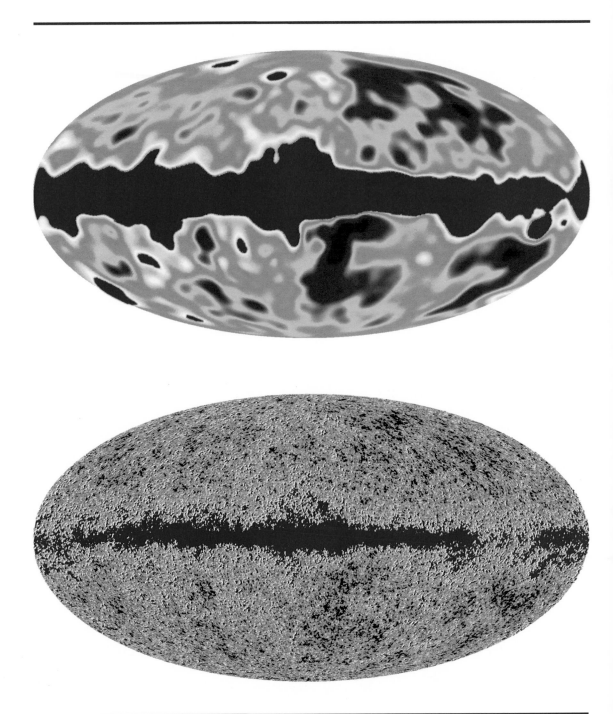

Hot Stuff. *A map of the cosmic microwave background obtained by the Cosmic Background Explorer (top), and a computer simulation of what the Microwave Anisotropy Probe may find (bottom). (Figures courtesy of NASA/MAP Science team, http://map.gsfc.nasa.gov.)*

Finite Math

The launch had been postponed again and again. Finally, on June 30, 2001, NASA's Microwave Anisotropy Probe (MAP) rocketed skyward from the Kennedy Space Center. Freelance geometer Jeff Weeks was delighted by the news. That's because the Canton, New York-based mathematician has a new-and-improved algorithm just waiting for data from the high-resolution satellite. Weeks has collaborated with scientists who will be analyzing MAP's map of the sky. And if all goes well, his algorithm may answer one of cosmology's most fundamental questions: What is the size and shape of the universe?

Theorists have pondered and debated the nature of space since antiquity. But only in the 20th century, with Einstein's general theory of relativity and astronomers' discovery of the expansion of the universe, has cosmology found its scientific footing. And only in recent years have cosmologists begun to think they may be able to tease precise answers out of astronomical observations.

Key to cosmologists' great expectations is the remnant whisper of primordial radiation known as the Cosmic Microwave Background. First detected by Bell Labs scientists Arno Penzias and Robert Wilson in the mid 1960s, the CMB is highly uniform:

Theorists have pondered and debated the nature of space since antiquity.

Jeff Weeks. *(Photo courtesy of the MacArthur Foundation.)*

No matter where you look in the sky, the spectrum of the microwave background is that of a theoretical black body radiating at a temperature of 2.73 degrees Kelvin, to within about a hundred thousandth of a degree. But those tiny departures are crucial. They echo the random fluctuations in density of the expanding early universe, which eventually spawned stars, galaxies, and astronomers.

Early observers couldn't see these fluctuations—it's like trying to eyeball bacteria—but data from the 1992 Cosmic Background Explorer (COBE) and more recent balloon-based measurements began to sharpen the picture. The CMB sky is now revealed as a patchwork of hot and cold spots (see Figure "Hot Stuff," page 32).

MAP will bring the picture into even sharper focus. "COBE had an angular resolution of 7 degrees. MAP has angular resolution of .2 degrees," says David Spergel, an astrophysicist at Princeton University and a member of MAP's science team. "So every place where COBE measured one number on the sky, MAP will measure a thousand."

In 1997, Spergel and colleagues Glenn Starkman at Case Western Reserve University and Neil Cornish, now at Montana State University, proposed testing whether the universe is finite by looking for "matched circles" in the fluctuations of the CMB. "When we look at the microwave background, we're basically looking out to the surface of a sphere," Spergel explains. If the universe is finite—and if it's small enough—then this "surface of last scattering," as it's called, must intersect itself (see Box, "Topology Makes the World Go Round," page 38). The intersections, if any, will show up as pairs of circles in different parts of the sky with precisely aligned temperature patterns.

Matched circles would be one signature of a finite universe. Researchers, including Spergel's group in the U.S. and a group in France headed by Jean-Pierre Luminet at Observatoire de Paris, are developing circle-spotting algorithms to crunch the data from MAP, hoping to find John Hancocks in the sky. That in itself is a major undertaking—in essence, their algorithms must examine all possible pairs of circles, to see if there are any matches. Only then does the question arise, what do the matches mean?

> **No matter where you look in the sky, the spectrum of the microwave background is that of a theoretical black body.**

That's where topology takes over.

The astronomers asked Weeks, an expert in topology, for help interpreting the matches they hope to find. They got more than they expected. "Jeff was the one who realized how you would use the circles to actually determine what was the underlying topology," Spergel says. Matched circles, Weeks saw, could be more than a signature; they could be a mathematical blueprint of the universe.

That's because each pair of matched circles provides a mathematical clue to the way space is stitched together. The easiest possibility to visualize is a "toroidal" universe slightly smaller than the surface of last scattering (see Figure 1). Such a universe can be thought of mathematically as a kind of magical box in which, when you exit through any wall, you immediately reenter through the opposite wall. There is, of course, no literal, physical wall, much as there's no edge to the Earth just because maps always show one. And much as U.S. map-makers conveniently place North America in the center, it's convenient to imagine Earth as being at the center of the box universe.

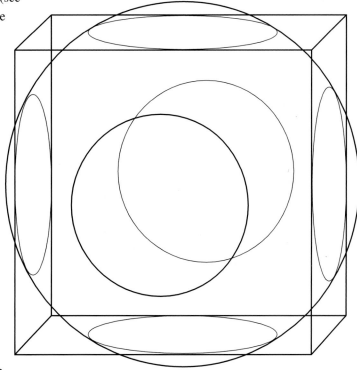

In this universe, the surface of last scattering intersects itself in circles on the walls. Since points on apparently opposite walls are in fact identical, this means that an observer on Earth would see the same circular patterns of cosmic radiation when looking in these three pairs of opposite directions. If MAP were to see three such pairs of matched circles, that would establish a toroidal geometry for the universe. (If the universe is toroidal but considerably smaller than the surface of last scattering,

Figure 1. *The topology of a toroidal universe slightly smaller than the surface of last scattering might reveal itself in three pairs of matched circles.*

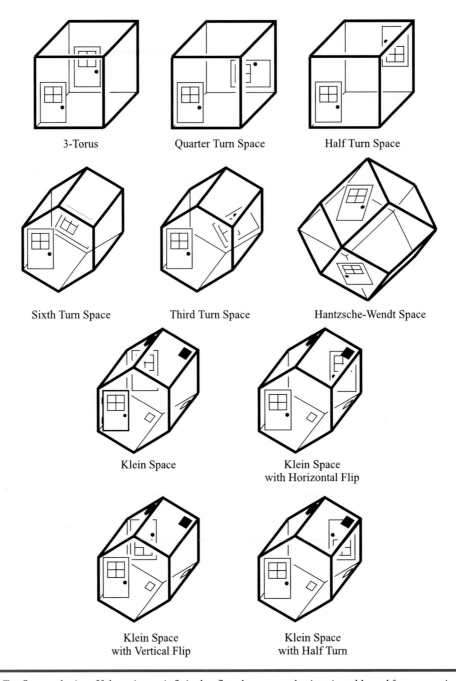

Figure 2. *Ten flat topologies. If the universe is finite but flat, there are only six orientable and four non-orientable possibilities. The doors in each figure indicate how a person exiting through one side would appear entering through another. In the first five orientable cases, the other sides are identified as they are for the torus. In the four non-orientable cases, the identification is indicated by the placement of small "windows." In the 12-sided Hantzsche–Wendt space, whose shape is a rhombic dodecahedron (see Figure 3 in "Nothing to Sphere But Sphere Itself," page 25), there are six pairs of identified sides (only one pair is shown). They are symmetrically pegged to the six points where four sides meet. (Figure courtesy of Adam Weeks Marano.)*

then the pattern of self-intersections becomes more complicated, but the math works out the same.) And if astronomers' best guess for the age of the universe is correct, that would establish the *size* of space as well.

Toroidality is only one possibility. It's one of ten different topologies for a "flat" universe (see Figure 2). And if the universe turns out to be curved—which is currently considered unlikely—then there are infinitely many more possibilities. Depending on whether the curvature is negative or positive, these alternative universes are what cosmologists traditionally call "open" or "closed" and mathematicians call "hyperbolic" or "spherical." (The near-uniformity of the CMB is evidence that the overall curvature of space is constant. There are, of course, local perturbations due to gravity, which are accounted for by the general theory of relativity.) Matched circles that correspond to a curved universe would contain an added bonus: The distance to the surface of last scattering (i.e., the age of the universe) and the radius of curvature of the universe would be related by a definite formula, giving a check on two otherwise independent numbers.

Weeks's original algorithm simply used matched circles to determine the "walls" of the universe. For example, if six pairs were found, the box would be a dodecahedron. Depending on how the pairs were oriented, the underlying geometry would be either spherical or hyperbolic. Whichever kind of box were found, purely mathematical considerations would then take over. In particular, if the geometry turned out to be hyperbolic, a computer program called SnapPea, which Weeks developed in the early 1990s, would finish the computation, determining the exact size and shape of space.

To work, though, this algorithm depended on having highly accurate data. But "real data are never as nice as mathematicans would like," Weeks notes. Not only will there be errors in the locations of observed circles, he notes, "but also there's the possibility of a few false matches creeping in, where two temperature patterns match to within some error tolerance just by chance. So you need some way of sifting out the false matches."

Weeks's new-and-improved algorithm does just that. It tolerates a considerable amount of error and uncertainty, and can even predict circles that the astronomers' circle-finding algorithms may have missed. The key new ingredient is group theory. Each pair of matched circles determines an element in a mathematical group.

Each pair of matched circles determines an element in a mathematical group.

Much like a crystal, the group necessarily has a rigid structure. This requirement locks in the geometry and corrects errors. The matched circles are something like regularly spaced trees in an orchard: The observations must reflect the regular spacing, but can also tell whether the orchard is on level ground or rolling terrain.

Because the new algorithm can cope with uncertainty, "we'll start taking a look as soon as any sort of data is available," Weeks says. "If there are some big circles out there just screaming to be observed, then we can get them with incomplete data. But if the incomplete data isn't good enough, then we'll look again with more complete data later on."

There is no guarantee, of course, that MAP will find any matched circles: The universe could in fact be infinite, or simply too large. Nevertheless, it's an exciting time to be a cosmologist, Spergel says: "In two years we could know that we live in a finite universe."

"In two years we could know that we live in a finite universe."

Topology Makes the World Go Round

The surface of last scattering is not a physical surface, like an eggshell or a balloon. It's a mathematically defined surface, consisting of points in the universe that currently are 15 billion (give or take a billion) light years from Earth—in other words, points from which the remnant microwave photons are only now arriving to be seen by astronomers. The self-intersections of this "virtual" surface are simply those points from which there are two straightline paths to the Earth of exactly the same length. For each such point, the two paths correspond to our looking in different directions in the sky—say toward Vega for one and Arcturus for the other.

How can there be two paths of the same length between two points (the Earth being the second point)? The answer is easiest to explain with 1- and 2-dimensional analogies. The basic idea is kin to the joking notion that if you could see all the way around the world, you'd be looking at the back of your head. A 1-dimensional, circular "universe" can be viewed mathematically as a periodically repeating straight line, which mathematicians call, appropriately enough, the "universal cover" of the circle. In 2 dimensions the simplest finite "universe" is a torus (more commonly called a donut). The universal cover of the torus is an infinitely repeating grid. If the "circle of last scattering" for such a universe is large enough, its repeating images intersect, in patterns ranging from simple to baroque (see Figure 3).

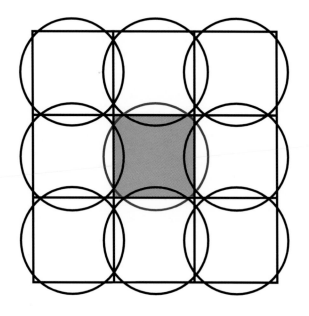

Figure 3. *In a 2-dimensional toroidal space, a "circle of last scattering" just slightly larger than its universe intersects itself in a simple pattern (top); a larger circle has a more intricate pattern of self-intersections (bottom). Something similar occurs in 3 dimensions, except that the self-intersections are circles rather than points.*

> **If two compact surfaces have the same number of holes, then they are topologically identical.**

Although a torus may look curved (and a real donut, of course, *is* curved!), its geometry is actually "flat," because its universal cover is simply the flat, Euclidean plane. But the torus is only one of many topological possibilities, most of which can only be realized using non-euclidean geometry. In 2 dimensions, every finite (technically called "compact") surface can be represented by a polygon with various pairs of sides identified. The torus, for example is a square (or, more generally, a parallelogram) with opposite sides identified. As it turns out, the same surface may be represented in more than one way (see Figure 4). But topogists have a simple way to tell when that happens: they can compute a number, called the "genus," which counts the number of holes in the surface. The torus, for example, has one hole, while the sphere has none. The genus is said to be an "invariant" of the surface—it stays the same no matter how the surface is deformed (picture a pliant balloon or underinflated inner tube).

A fundamental theorem in topology states that if two compact surfaces have the same number of holes, then they are topologically identical. In other words, each surface is characterized by a single invariant, namely its genus. (To be precise, this only applies to "orientable" surfaces. The theory also treats non-orientable surfaces, such as the Klein bottle, a near relative of the famous Möbius strip.)

Things are much more complicated in 3 dimensions (and even more so in 4-d). Finite spaces can still be represented geometrically, in this case by polyhedra with various pairs of faces identified. But mathematicians have no sure way of telling whether or not two spaces, represented by different polyhedra, are topologically the same. No single invariant, such as genus, is known to be different for different spaces.

There is, however, a conjectured way to classify 3-d spaces. Known as the Thurston Geometrization conjecture, this classification says, roughly speaking, that every 3-d space can be uniquely decomposed into pieces, each of which has a prescribed geometric structure. Moreover, in most cases ("most" having a precise mathematical meaning) the geometry is hyperbolic. The conjecture was formulated in the 1970s by William Thurston, now at the University of California at Davis, who proved that this approach accounts for a large class of spaces known as Haken manifolds. Mathematicians are confident that the Geometrization conjecture is correct—like the Taniyama–Shimura conjecture (see "New Heights for Number Theory," page 2), it has been checked in a great many cases and is seen as a powerfully unifying theory—but a proof seems to be a long ways off. On the other hand, things that appear distant in one direction may actually be quite close.

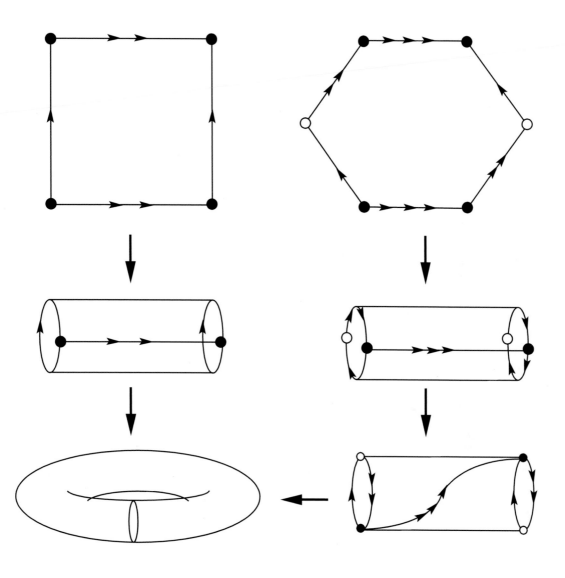

Figure 4. *Two different polygons can produce the same topology.*

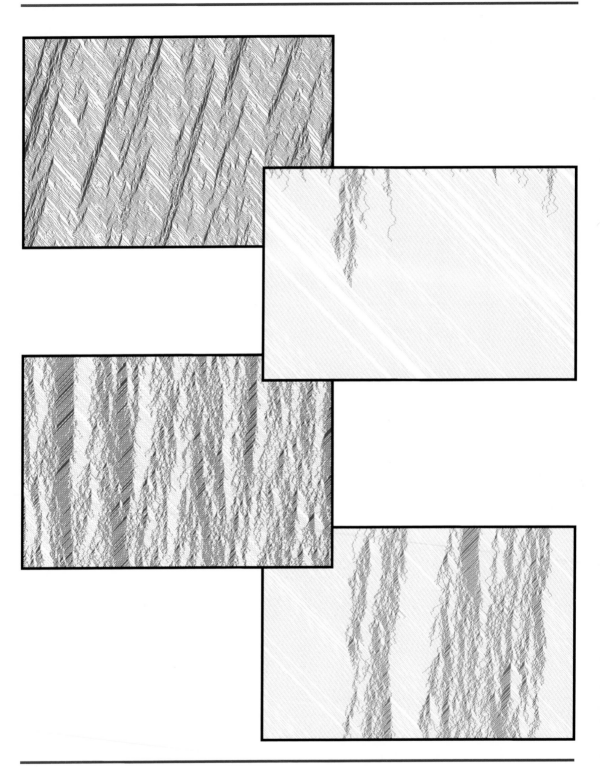

Traffic Flow. *Simple mathematical models can capture some of the complex dynamics that occur when cars crowd onto a freeway. (Figures courtesy of David Griffeath.)*

The Mathematics of Traffic Jams

I t's one of the mysteries of modern life: You're cruising along a
freeway when, suddenly, traffic grinds to a halt. You inch along
for fifteen minutes until, just as suddenly, the road ahead is clear
again, and you're back at the speed limit (or a little above)—until
you hit another jam.

What causes a stretch of highway to become a temporary parking lot? A new theory using mathematical models based on cellular automata is giving insight into the question. Computer simulations based on the models mimic many of the annoying features of
real traffic jams. Among the implications of the theory: Once the
density of traffic reaches a certain critical value, jams become an
unavoidable, self-sustaining feature, even though the traffic could,
in principle, be moving snarl-free.

Researchers have long looked for mathematical explanations of
traffic patterns. One of the early efforts came in the 1950s. James
Lighthill and Gerald Whitham at the University of Manchester, two
experts in fluid dynamics, reasoned that the equations used to
describe the flow of water might also describe the flow of traffic.
In fluid dynamics, these are a set of partial differential equations
called the Navier–Stokes equations. They are derived from physical laws such as the conservation of mass and momentum. Similar
concepts apply to traffic. In particular, the density of traffic along
a certain stretch of road can change in time only if more traffic
flows in than out. Lighthill and Whitham found that traffic jams
can be likened to shock waves: Both exhibit sharp discontinuities
in density, such as occur in a sonic boom.

One obvious drawback of the fluid-flow approach is that traffic
is clearly not a continuum, but consists of discrete "particles." (Of
course this is also true of fluids, but only at the molecular scale,
which is nanoscopically smaller.) An alternative approach is
known as car-following theory. Here two variables are associated
with each car: its own speed and its distance to the car immediately ahead. Each car then adjusts its speed—slamming on the brakes
or putting the pedal to the metal—based on a calculation with these
variables. For example, each driver might try to follow the two-
second rule, accelerating when the next car is more than two seconds ahead at current speed, and braking when the gap is less. A
car-following model determines a dynamical system that can be

**What causes a stretch
of highway to become a
temporary parking lot?**

Kai Nagel. *(Photo courtesy of Kai Nagel.)*

analyzed mathematically or simulated on a computer in search of steady states (all vehicles eventually driving at the posted speed—yeah, right!), periodic (stop-and-go) patterns, or even chaos.

A more recent approach, known as particle hopping, is based on the mathematical theory of cellular automata (see "Cellular Automata Offer New Outlook on Life, the Universe, and Everything," *What's Happening in the Mathematical Sciences,*" Volume 3). At first blush, this approach sounds extremely simplified. Instead of treating traffic as continuously in motion, a particle hopping model views everything as proceeding in discrete steps. At time step k, the position of car i is an integer $n_i(k)$; its velocity is another integer, representing the amount to move forward at the next time step—i.e., $n_i(k+1) = n_i(k) + v_i(k)$. In most particle hopping models there is a probabilistic element in the velocity, mimicking the random variations that occur when drivers accelerate or brake.

The particle hopping approach to traffic was introduced in the early 1990s. A leading theorist is Kai Nagel, a physicist now at the Institute for Scientific Computing at the Swiss Federal Institute of Technology (ETH) in Zurich. Discrete models have long been used to study such phenomena in statistical physics as phase transitions: the abrupt changes of state that occur, for example, when ice melts at precisely 32 degrees Fahrenheit, or when water boils at 212 degrees (see "Ising on the Cake," page 88). Nagel and colleagues believe that traffic exhibits similar phase transitions. Roughly speaking, when there are few cars on the road, they move freely, much like molecules in a gas; at a certain density, the flow becomes "synchronized"—that is, cars are forced to move at more or less the same speed, like molecules in liquid; and at higher densities, cars find themselves frozen in place like a solid.

Making quantitative sense of these analogies requires mathematical finesse. In 1992, Nagel and Michael Schreckenberg at the University of Duisburg, Germany, introduced the first particle hopping model for traffic. In their model, each car i keeps track of its current velocity v_i and the distance d_i to the car ahead of it. The velocity is allowed to be any integer between 0 and a "speed limit," v_{max}. (In computer simulations, the speed limit was 5.) At each time step, each car going under the speed limit initially increases its speed by 1 ($v_i \rightarrow v_i + 1$); however, any car going fast enough to potentially crash into the car ahead of it—that is, any car whose velocity v_i is greater than or equal to d_i—*decelerates* to $v_i = d_i - 1$; each car then randomly decelerates ($v_i \rightarrow v_i - 1$, provided $v_i > 0$) with probability p; and finally each car moves forward the amount indicated by its speed.

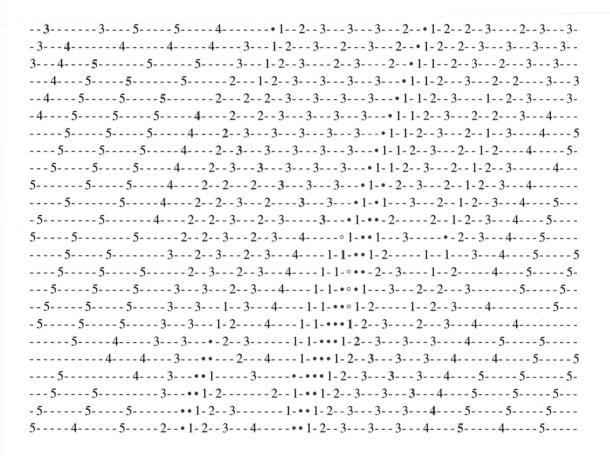

Figure 1. *Particle hopping simulation. When 20 cars travel on an 80-space track, jams tend to occur. Numbers represent the cars' current speeds (with teal dots for jammed cars whose speed is 0). This simulation used a deceleration probability $p = 0.1$. One car is highlighted as it moves with traffic. (When cars "exit" on the right, they re-enter on the left.) A persistent traffic jam drifts backwards through traffic.*

Simulations with Nagel and Schreckenberg's model show some of the classic features of traffic jams (see Figure 1, page 45). For example, tie-ups tend to travel backwards: New cars pile up at the rear of a jam as those at the front are finally freed to accelerate. Most important was the observation of a "critical density" below which jams are only transient, but above which they are a permanent feature.

Nagel and his colleagues have investigated many different particle hopping models, including models with more than one lane of traffic. As a researcher at Los Alamos National Laboratory from 1995 to 1999, Nagel worked on TRANSIMS, a computer program for highly realistic traffic simulations (see Box, "Commuting and Computing"). The models have identified a crucial factor in perpetuating traffic jams. Researchers call it the "slow to start" rule: Cars emerging from a jam speed up less quickly than conditions would permit. This accords with how drivers actually behave: How often, while trapped in a backup that finally starts to move, have you let the car in front of you get several car lengths ahead before budging yourself?

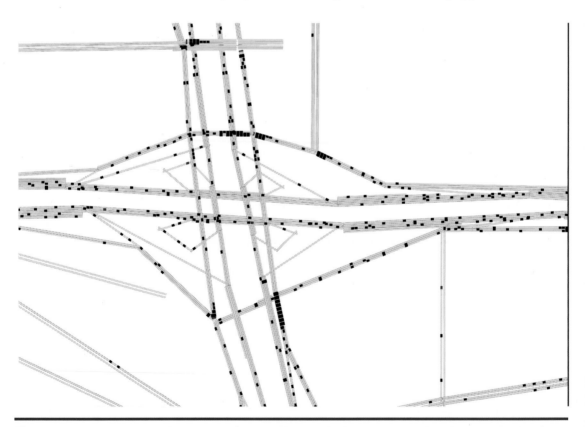

Figure 2. *Zoomed-in snapshot of a micro-simulation in TRANSIMS. (Figure courtesy of Kai Nagel, Richard Beckman, and Christopher Barrett.)*

Commuting and Computing

Traffic jam researchers are counting on simple mathematical models to provide fundamental insights into the nature of vehicular flow. At the same time, others are developing elaborate computer models to simulate realistic traffic patterns in real cities. One of the biggest projects is TRANSIMS (Transportation Analysis and Simulation System), developed at Los Alamos National Laboratory with funding from the Department of Transportation's Federal Highway Administration.

Headed by Christopher Barrett and Laron Smith, the TRANSIMS team has done detailed simulations of transportation in Albuquerque, Dallas–Fort Worth, and Portland, Oregon. Modeling the flow of cars is just one component. "Vehicular traffic is important, but we have to answer questions like what happens to traffic and daily activity structures if bus tickets get a lot cheaper?" Barrett says. "Our customers want insight into the entire transportation system."

To give such insight, TRANSIMS takes an "agent-based" point of view: It follows *people*, not vehicles. The program starts with demographic data, from which it creates a synthetic population: a virtual world "like SIMCITY, but much more realistic," says Kai Nagel, who worked on the project in the 1990s. The computer creates a schedule of activities for each individual, from sleeping and eating to working and shopping. It then has each agent plan how to get from place to place, after which it runs a detailed micro-simulation of the resulting traffic (see Figure 2).

TRANSIMS can be used to analyze everything from car pollution to the impact on traffic of a new shopping center. Because it's so finely detailed, it can be used for epidemiological studies (e.g., how will next year's flu spread from person to person?), or by cell phone companies wanting to know where their customers are likely to be. "A lot of things are driven by population mobility," Barrett notes. Similarly, population mobility is affected by the way a city is laid out. In particular, Barrett says, "if the city is poorly designed, then the traffic will create jams."

TRANSIMS can be used to analyze everything from car pollution to the impact on traffic of a new shopping center.

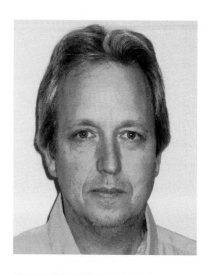

David Griffeath. *(Photo courtesy of David Griffeath.)*

Another key insight is that traffic jams are "self-organizing" phenomena: They occur under a wide range of conditions, and their basic nature does not depend on the specifics of the model (although, of course, their detailed properties do). Mathematicians David Griffeath at the University of Wisconsin and Lawrence Gray at the University of Minnesota conclude from that insight that traffic jams can be studied with an extremely simple model.

The advantage of doing so, Griffeath says, is that simple models yield to mathematical analysis; the researchers have been able, for example, to prove the existence of a critical density for traffic jams in their model. In fact, Griffeath says, their model offers the first rigorous mathematical definition of what a traffic tie-up actually is.

In Griffeath and Gray's model, each car looks two spaces ahead and one space back. If the space immediately ahead is already occupied, the car does nothing. (This is the "no crash" clause, which all particle hopping models obey. Car crashes do happen in the real world, of course, and they create massive traffic tie-ups, but most traffic jams are accident-free.) If the space ahead is open, the car determines which of four "states" it's in: *accelerating* (a car behind but none two spaces ahead), *braking* (no one behind but someone two ahead), *congested* (both places occupied), or *driving* (neither occupied).

The intuition behind Griffeath and Gray's nomenclature is that the first state typically occurs when a car is emerging from a jam (so it should be accelerating), the second when it's entering a jam (so it should be braking), the third when it's in a jam (i.e., congested), and the fourth when it's on open highway (so it should simply

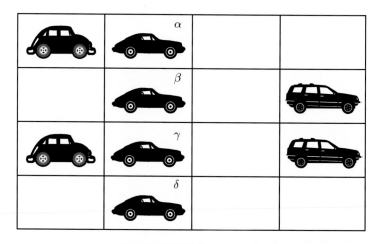

Figure 3. *Four states of traffic with a "no crash" clause: accelerating, braking, congested, and driving.*

be driving). The rest of the model is equally simple: The car moves forward with probability α, β, γ, or δ, depending on which state it's in (see Figure 3). In particular, the slow-to-start rule is simply the inequality $\alpha < \delta$: A car is less likely to move forward when it finally gets to the front of a jam than when it's on open highway.

In computer simulations, Griffeath and Gray set up a virtual highway, with hundreds or thousands of spaces for cars to occupy. To keep the number of cars constant, the highway runs in a circle, like a racetrack. (On the screen, the track is a horizontal line. When cars reach the right end, they re-emerge at the left end.) After setting values for α, β, γ, and δ and picking a density of traffic (traditionally denoted ρ), they display the resulting patterns as a cascade of pixels: black for occupied spaces and white for unoccupied. (Actually their program uses a larger palette, color-coding cars according to the states they're in and whether they moved forward or not.) Jams show up as dark streaks, generally drifting leftward as the cascade proceeds from top to bottom (see Figure 4).

That leftward drift corresponds to the familiar, backward-propagating property of real traffic jams. Griffeath and Gray have analyzed conditions under which jams in their model occur and persist. In particular, they have examined what they call the case of "symmetric cruise control," in which $\gamma = \delta = 1$.

Setting $\delta = 1$ is sensible enough. It just says that drivers go as fast as they can on open road—just as "cruise control" implies. Setting $\gamma = 1$, however, sounds paradoxical, since it implies that

Larry Gray. *(Photo courtesy of Larry Gray.)*

Figure 4. *Four hundred time steps of 240 cars ($\rho = .4$) on a 600-space track, with $(\alpha, \beta, \gamma, \delta) = (.5, .4, .3, .9)$. Cars move left to right, but jams travel backwards. (Figure courtesy of David Griffeath.)*

drivers also go as fast as they can in congested conditions. But in a sense drivers *do* go as fast as they can in congested conditions—it's just that they don't go very far before they have to slow down again. In Griffeath and Gray's model, a car that experiences congestion at one time step will either brake on the next time step or not move at all, depending on whether the car two spaces ahead of it moved (see Figure 5).

The real reason for setting $\gamma = \delta$, Griffeath explains, is that doing so gives the model a special symmetry: The empty spaces can be viewed as "anti-cars," which move from right to left in precisely the same way the cars move left to right. This view is similar to the use of electron "holes" in the study of transistors. In the traffic model, particle hopping takes place only where cars and anti-cars meet head on, and the symmetry is to interchange their roles (see Figure 6). Griffeath and Gray don't expect traffic engineers to concern themselves too much with anti-cars (imagine reports of anti-car pile-ups during the midday anti-rush hour), but the symmetry helped them discover the self-organizing properties of the model.

If the density is low enough, then no matter how the cars are initially spaced—in Griffeath and Gray's simulations they are arranged randomly—they will eventually settle down with at least two spaces between consecutive cars; then cruise control takes over and the traffic moves at maximal speed thereafter (see Figure 7, page 52). The "throughput" of the system, defined as the average rate at which cars pass a given point on the roadway, is then simply the density ρ.

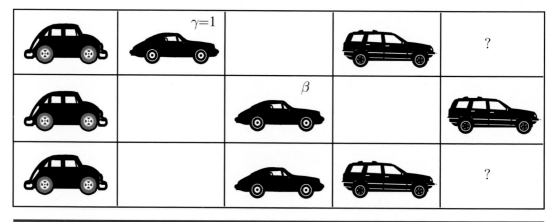

Figure 5. *In the case of symmetric cruise control, a car experiencing congestion at one time step (top row) will either brake on the next (middle) or not move at all (bottom), depending on the car ahead. Question marks indicate the driver does not know whether those spaces are occupied.*

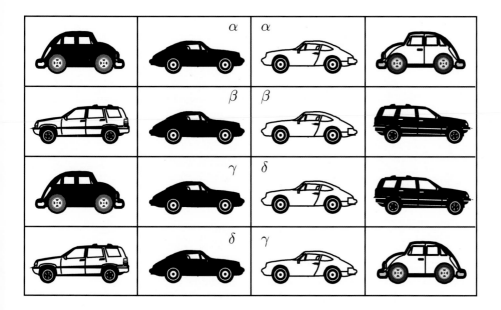

Figure 6. *Cars and anti-cars.*

Such "free flow" can, of course, be engineered for densities all the way up to $\rho = 1/3$ by initially spacing cars so that they start out—and stay—in cruise control. And in a finite setting, such as a racetrack simulation, cars will again, in principle, stumble into free flow, although the number of time steps it takes to do so becomes astronomically large. (The waiting time resembles that needed to get 100 heads in a row when tossing a coin repeatedly.) But in the theoretical setting—infinitely many spaces, each initially occupied with probability ρ—the probability that any car will ever find itself permanently in cruise control drops precipitously from 1 to 0 at a critical density ρ_* less than 1/3.

By the car–anti-car symmetry, similar results apply for densities between 2/3 and 1. The most important densities ρ are those between ρ_* and 1/2. In simulations with densities near 1/2, the entire racetrack becomes permanently jammed, with cars only occasionally able to cruise more than a few spaces (see Figure 8, next page). But just below another critical density ρ^*, something new happens: The traffic flow separates into two components, with parts of the racetrack jammed and other parts in free flow (see Figure 9, page 53).

This "clustering" behavior occurs for *all* densities between ρ_* and ρ^*. What's more, Griffeath and Gray have found, clustering occurs in such a way that the density of the jammed traffic is, on average, exactly ρ^*, while that of the free-flowing traffic is ρ_*. In

Figure 7. *Four hundred time steps of 168 cars ($\rho = .28$) on a 600-space track, with $(\alpha, \beta, \gamma, \delta) = (.6, .6, 1, 1)$. Jams caused by the initial random placement of the cars disappear. (Figure courtesy of David Griffeath.)*

Figure 8. *Four hundred time steps of 288 cars ($\rho = .48$) on a 600-space track, with $(\alpha, \beta, \gamma, \delta) = (.6, .6, 1, 1)$. Gridlock is pervasive and persistent. (Figure courtesy of David Griffeath.)*

other words, their model exhibits the same kind of self-organized criticality that physicists have found in many other phenomena.

Since the overall density ρ is fixed, a simple calculation says that the fraction of racetrack with free-flow conditions is $(\rho^* - \rho)/(\rho^* - \rho_*)$, while jams occupy the remaining fraction, $(\rho - \rho_*)/(\rho^* - \rho_*)$. In the finite setting of simulations, most jams coalesce, until there is just one big jam. In the theoretical, infinite setting, however, the separation persists at all scales: The overall fractions remain the same, but there are arbitrarily large jams and arbitrarily large stretches of free-flow.

Real traffic, of course, is not infinite, nor does it simply go round and round a racetrack—no matter how we may feel doing our daily commute. Griffeath and Gray's theorems and simulations don't translate directly into highway design. But the insights theorists have discovered can guide the development of more realistic computer models, which may someday reduce traffic crunches—and crashes. We might even hope for a mathematical explanation of the rush-hour paradox: How waiting ten minutes before leaving home gets you to the office five minutes early.

> **In the finite setting of simulations, most jams coalesce, until there is just one big jam.**

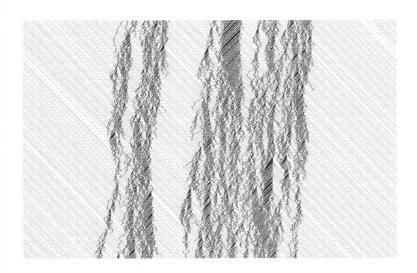

Figure 9. *Four hundred time steps of 228 cars ($\rho = .38$) on a 600-space track, with $(\alpha, \beta, \gamma, \delta) = (.6, .6, 1, 1)$. Traffic separates into jammed and free-flow regions. (Figure courtesy of David Griffeath.)*

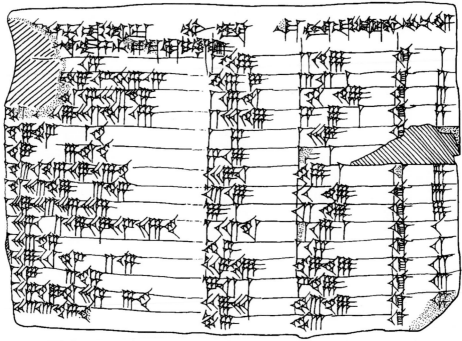

Plimpton 322. *(Photo from* Mathematical Cuneiform Tablets, *Otto Neugebauer and Abraham Sachs, 1945, courtesy of the American Oriental Society; drawing by Eleanor Robson.)*

Rewriting History

Most mathematicians focus on making new mathematical discoveries and finding new applications of mathematics. But some scholars look backward, at work done long ago. Even there, new discoveries are possible. Recent historical research by Eleanor Robson, an Oriental scholar at the University of Oxford, has shed new light on an ancient work of algebra—a cuneiform tablet known as Plimpton 322.

Plimpton 322 is one of thousands of clay tablets dating back to the Old Babylonian period in Mesopotamia nearly 4000 years ago. It resides in the George Arthur Plimpton collection at Columbia University. The tablet was purchased by Plimpton in 1923, from Edgar J. Banks, who said it came from a location near the ancient city of Larsa (modern Tell Senkereh) in Iraq. Robson estimates that Plimpton 322 was created sometime in the six decades before Larsa fell to Hammurabi of Babylon in 1762 BCE.

At 12.7×8.8 cm, Plimpton 322 is about the size of a cheap pocket calculator. (It was originally larger: There is a clean break along the left edge. Remnants of glue suggest the missing piece might still be in a drawer somewhere, but Robson suspects that Banks simply removed an unrelated fragment of tablet that a less scrupulous antiquities dealer had glued on to make the piece look complete.) Both sides are ruled like notebook paper. The back side is otherwise blank, but the front side has 15 lines of numbers arranged in four columns, with some labeling across the top.

The rightmost column, labeled "its name," contains only the numbers 1 through 15. It's the other three columns that make the tablet fascinating to mathematicians.

Plimpton 322 was originally assumed to be just another Babylonian ledgerbook, a kind of Sumerian spreadsheet. (The Babylonians were avid record keepers. Many of their tablets list acreages of wheat and quantities of livestock. Others are tax forms that would have done the IRS proud.) But in the early 1940s, Otto Neugebauer, an historian of ancient science at Brown University, and his assistant Abraham Sachs found otherwise. They recognized the entries as, in effect, Pythagorean triples: integer solutions of the equation $a^2 + b^2 = c^2$.

Pythagorean triples are most closely associated with right triangles. They also make for nice solutions to "reciprocal pair"

Edgar J. Banks. (*Photo from* Bismaya, or the Lost City of Adab, *by Edgar J. Banks (1908), provided courtesy of The University of Chicago Library's Electronic Open Stacks, www.lib.uchicago.edu/eos.*)

(1).9834	119	169	1
(1).9416	3367	11521	2*
(1).9188	4601	6649	3
(1).8862	12709	18541	4
(1).8150	65	97	5
(1).7852	319	481	6
(1).7200	2291	3541	7
(1).6928	799	1249	8*
(1).6427	541	769	9*
(1).5861	4961	8161	10
(1).5625	45	75	11
(1).4894	1679	2929	12
(1).4500	25921	289	13*
(1).4302	1771	3229	14
(1).3872	56	53	15*
(1).3692	175	337	16

Figure 1. *Decimal conversion of P322. The first column is rounded to 4 decimal places. Row 16 does not appear on the tablet. The sharp-eyed reader may notice mistakes (made by the author of Plimpton 322) in the rows marked with asterisks.*

equations of the form $x - 1/x = 2b/a$ and $x + 1/x = 2c/a$. (In each case, $x = (b + c)/a$ is a solution.) Old Babylonian scribes were trained on both kinds of problem, although of course they didn't use modern, algebraic notation—or refer to Pythagoras, who wouldn't appear for another thousand years!

The numbers in the middle two columns of Plimpton 322 record the short side b (with $a > b$) and hypotenuse c for a list of right triangles. The leftmost column contains the ratios $(c/a)^2$ (or possibly $(b/a)^2$, depending on whether the entries there do or don't start with a "1"—the break along the left edge makes it ambiguous). This column also provides the organizing principle: The ratios decrease from top to bottom, beginning (in decimal equivalent) at (1).9834 and ending at (1).3872 (see Figure 1).

Ever since Neugebauer's discovery, mathematicians have had a field day interpreting Plimpton 322 and speculating on how it was composed. The most popular explanation, advanced by Neugebauer, is that the scribe responsible for Plimpton 322 knew that setting $a = 2pq$, $b = p^2 - q^2$, and $c = p^2 + q^2$, with integers p and q (with $p > q$) would produce Pythagorean triples. Armed with this knowledge, he had, in effect, plugged in all the small values of p and q for which a has an exact sexagesimal (base 60) reciprocal—i.e., values with prime divisors 2, 3, and 5. Doing so makes the numbers in the first column exact. (In our modern decimal system, the only fractions that can be written as exact decimals are those with 2's and 5's in the denominator.) The tablet contains the results corresponding to right triangles with base angle ranging from $45°$ down to around $30°$ (omitting, however, the pair $p = 16$, $q = 9$—see Figure 2).

That ingenious

Eleanor Robson. *(Photo courtesy of Eleanor Robson.)*

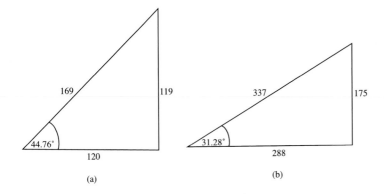

(a)   (b)

Figure 2. *Right triangles corresponding to the first and "last" entry of P322. The latter, which corresponds to $p = 16$, $q = 9$, is not actually on the tablet. The angle for the actual last line is $31.89°$.*

explanation implies a flattering view of the creator of Plimpton 322: a mathematical prodigy, doing original research in number theory and trigonometry. Robson favors a more mundane alternative. The tablet, she believes, was created as a teacher's aid, designed for generating problems involving right triangles and reciprocal pairs. Robson has marshalled her arguments in a recent article in *Historia Mathematica*.

A reciprocal pair are two numbers whose product is a power of 60. (Lacking a symbol for zero, the Old Babylonian scribes would write any such power simply as "1.") There are several examples of cuneiform tablets with school problems like "A number and its reciprocal differ by 7. What is the number?" The solution proceeds by a kind of "cut and paste" geometry (see Figure 3, next page): Divide 7 by 2, getting 3;30 (i.e., $3 + 30/60$), square 3;30 to get 12;15, "add" 1 to get 1;12;15, and find the square root of that, which is 8;30 (You can check this by computing

$$(8;30)^2 = (8 \times 60 + 30)^2 = 510^2 = 260100$$
$$= 1 \times 60^3 + 12 \times 60^2 + 15 \times 60 = 1;12;15.$$

The Babylonians were keenly interested in squares and square roots.) Finally, add and subtract 3;30 to 8;30, getting 12 and 5, respectively. These numbers are reciprocals, and they differ by 7.

In modern terms, if x and y are a reciprocal pair (that is, $xy = 1$), then $1 + ((x - y)/2)^2 = ((x + y)/2)^2$. This implies that $a = 1, b = (x - y)/2$, and $c = (x + y)/2$ are a Pythagorean triple,

The *p/q* theory fails to account for many of the features of the tablet.

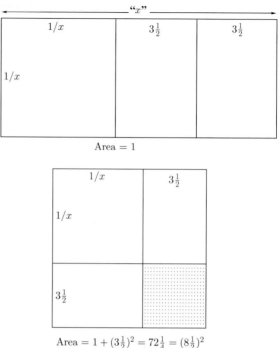

Figure 3. *Cut and paste mathematics. In working with reciprocal pairs, sometimes "1" means 60.*

once you clear out the denominators in b and c. To be a reciprocal pair in the Old Babylonian sense, x and y must have the form p/q and q/p with 2, 3, and 5 as the only prime divisors of p and q. Thus, clearing denominators leads to $a = 2pq$, $b = p^2 - q^2$, and $c = p^2 + q^2$. In other words, the p/q theory and the reciprocal pairs explanation are mathematically equivalent from a modern viewpoint.

But the author of Plimpton 322 did not have a modern viewpoint. According to Robson, the p/q theory fails to account for many of the features of the tablet, including that fact that it records values of $(c/a)^2$ instead of a. The reciprocal pair explanation, she says, makes more sense in light of what's been learned about Old Babylonian tablets in the last half century.

One key is the label for the first column. Neugebauer and Sachs rendered it as "The *takiltum* of the diagonal which has been subtracted such that the width...," leaving *takiltum* untranslated and the label unfinished, because part of it near the end is unreadable. ("Diagonal" means "hypotenuse," since right triangles arise by cutting a rectangle diagonally in half. "Width" and "short side" are also synonymous.) Subsequent scholars, observing the use of *takiltum* in other mathematical tablets, determined that it refers to a "helping" or "holding" number. With that meaning and an educated guess for what makes grammatical sense (and also fits physically) in the unreadable and damaged portions, Robson offers a new translation: "The holding-square of the diagonal from which 1 is torn out, so that the short side comes up." That reading, she says, aligns well with the Old Babylonian approach to solving reciprocal-pair-type problems and with other mathematical tablets of the time. So it seems that the author of Plimpton 322 was no lone genius—but he was probably a very good teacher.

Take Home Assignment

Thanks to mathematicians such as René Descartes and Pierre de Fermat, many of the problems that the ancients struggled with are a virtual snap today. Finding Pythagorean triples, for example, is a simple exercise in algebraic geometry, since $a^2 + b^2 = c^2$ (with $c \neq 0$) if and only if $(a/c, b/c)$ is a point on the "unit circle," which is defined by the equation $x^2 + y^2 = 1$. But a point (x, y) on the unit circle has rational coordinates if and only if the line connecting it to $(-1, 0)$ has rational slope (see Figure 4a). If the slope is taken as p/q, then a little algebra leads to $x = (q^2 - p^2)/(q^2 + p^2)$ and $y = 2pq/(q^2 + p^2)$, giving the now-familiar formula for Pythagorean triples.

A similar approach works for problems like "Find three squares in arithmetic progression"—that is, a^2, b^2, c^2 with $b^2 - a^2 = c^2 - b^2$. Solutions correspond to rational points $(a/b, c/b)$ on the circle $x^2 + y^2 = 2$. In this case, a point (x, y) has rational coordinates if and only if the line connecting it to $(-1, -1)$ has rational slope (see Figure 4b). This time the algebra leads to $x = (q^2 + 2pq - p^2)/(q^2 + p^2)$ and $y = (p^2 + 2pq - q^2)/(q^2 + p^2)$, which gives $(a, b, c) = (q^2 + 2pq - p^2, q^2 + p^2, p^2 + 2pq - q^2)$ for numbers whose squares are in arithmetic progression. The values $(p, q) = (2, 1)$, for example, give $(1, 5, 7)$, corresponding to squares 1, 25, and 49.

All this still works fairly easily if you generalize to Pythagorean "quadruples," in which three squares sum to a square: $a^2 + b^2 + c^2 = d^2$. But if you go looking for four squares in arithmetic progression, the going gets a little tricky. Try it!

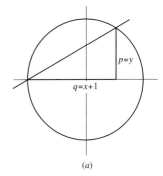

(a)

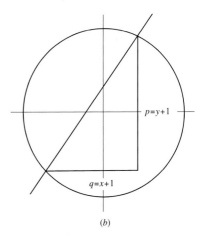

(b)

Figure 4. *Rational points on the circle $x^2 + y^2 = 1$ (a) and $x^2 + y^2 = 2$ (b).*

Goldilocks and the Three Networks

Goldilocks had barely escaped with her life from the home of the three bears. You'd think she would have learned not to help herself to the belongings of others. But no.

When she was done running and had caught her breath again, Goldilocks found herself outside the home of the three spiders. Without so much as a knock on the door, she went inside and sought a spidery hammock to nap in after her long run.

"Oh, this web is much too random," she said of the first. "Parts of me will be comfortably supported, but other parts hardly at all."

"Oh, this web is much too latticelike," she said of the second. "A couple of broken strands, and the whole thing might fall apart. Besides, I don't want to wake up looking like a checkerboard."

"Oh, this web is just right!" she said of the third. So in she climbed and asleep she fell, her goldeny locks entwining with the silvery, silken spider strands. *(To be continued)*

Just Right. *Researchers have found that "small world" networks are a good fit for many problems.*

It's a Small, Big, Small, Big World

Mathematicians often find it useful to look at extremes. In the theory of networks, which concerns the dynamics of systems with interconnected parts, the extremes are completely random networks at one end and highly regular networks at the other end. But theorists are coming to appreciate networks that fall between the two extremes.

The new subject's name, "small world networks," recalls the buzz phrase "It's a small world," often heard when strangers discover a common link between them. Two airline passengers, for instance, might find that one's uncle works in the same office as the other's wife. A small world network is an amalgam of local clusters, such as cliques of friends or officeworkers, and random long-range links, as might occur through a high-school exchange program or when a husband and wife work for different companies. Among other applications, the theory may help sociologists track how information and ideas, including fads and fashions, spread through a community. It may also help epidemiologists model infectious diseases more accurately. And it's not just networks of people that exhibit small-world effects: The nervous system may be wired along the same principle.

The small-world effect, sometimes called six degrees of separation, was famously studied by the psychologist Stanley Milgram at Yale University. In a 1967 article in *Pyschology Today*, Milgram described an experiment in which people in Omaha, Nebraska, were given a letter intended for delivery to a stranger in Massachusetts, and asked to pass it along to someone they thought might know the intended recipient, with instructions for that person to do the same. Only about 20% of the letters ever reached the "target." For those that did, though, the surprise was how quickly they arrived: It rarely took more than 6 intermediate steps (see Figure 1).

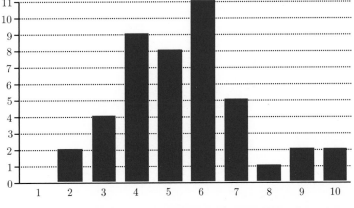

Figure 1. *In a classic study conducted by psychologist Stanley Milgram, 44 out of 160 letters made it from Nebraska to Massachusetts. The letters that arrived took between 2 and 10 intermediaries, with a peak at 6. (From data in Milgram, "The Small-World Problem," Psychology Today, May, 1967.)*

Steven Strogatz. *(Photo courtesy of Steven Strogatz.)*

Duncan Watts. *(Photo courtesy of Myrna Suárez/Twin b. Photography.)*

The mathematical study of small world networks was kickstarted in 1998, in a paper published in *Nature* by Duncan Watts and Steven Strogatz at Cornell University. They looked at idealized networks consisting of points, or "nodes," around a circle, with each node initially connected to just a few of its neighbors (see Figure 2a). If each of N points is connected to k nearby neighbors, then it's easy to see that the maximum "distance" between two nodes is roughly N/k. A slightly more complicated calculation shows that the *average* distance between two randomly chosen points is approximately $N/2k$. That's a lot of links to line up: If N is 200 million and k is, say 100, the average distance is a million.

Next, they "rewired" this regular network completely at random, using the same total number of connections, so that each node is, on average, connected to k other nodes (see Figure 2c). But the average distance between nodes now drops to around $\log(N)/\log(k)$. This result (which is not so easy to prove) typifies the small world phenomenon: If each of, say, 200 million people knows 100 other randomly chosen people, then the average pair of strangers are linked by a chain of three intermediaries $(\log(2 \times 10^8)/\log(100) = 4.15)$.

But, of course, people don't make friends completely at random. The people you know tend to know each other. Watts and Strogatz call this "clustering." Each node in a network has a cluster value, defined by counting the connections among the given node's "friends." If a node has k neighbors, there can be at most $k(k-1)/2$ such connections; the node's cluster value is the fraction of these possibilities that actually occur. The network's "clustering coefficient," C, is the average of all nodes' cluster values.

The clustering coefficient in completely random networks is around k/N. In Watts and Strogatz's highly regular networks, on the other hand, it hovers around the constant value 3/4. (The precise value in their model is

$\frac{3}{4}(1 - \frac{1}{k-1})$. The key point is that it is independent of N.) So—if you look only at the extremes—it would seem that tight local clusters entail long average distances, and that reducing average distance requires abandoning local clustering.

But extremes can be deceptive. In a small world, the clustering coefficient stays near 3/4, while the average distance between nodes is proportional to $\log(N)$ rather than to N. This is exactly what Watts and Strogatz observed when, instead of randomly rewiring *all* connections in their highly regular network, they rewired only a small fraction of the connections (see Figure 2b).

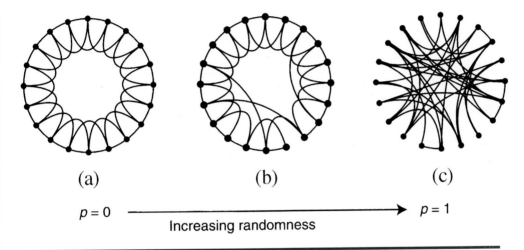

(a)   (b)   (c)

$p = 0$ ⟶ $p = 1$

Increasing randomness

Figure 2. *Small world networks lie between highly regular and thoroughly random networks. Here the probability of rewiring each connection increases from 0 (a) to 0.1 (b) to 1 (c). (Reprinted with permission from* Small Worlds, *by Duncan J. Watts (1999), Princeton University Press.)*

The new method of rewiring is to consider each connection in turn, rewiring one end of it with probability p. This leads to a continuum of networks, from highly regular ($p = 0$) to completely random ($p = 1$). The small-world networks are in between.

In computer simulations of networks with $N = 1000$ nodes and $k = 10$ neighbors per node, Watts and Strogatz found that the clustering coefficient stays near its peak value, 3/4, for probabilities up to 1%, and remains relatively large even for values of p above 10%. The average distance, on the other hand, drops rapidly from its "big-world" value of $N/2k = 50$ to the small-world value $\log(N)/\log(k) = 3$ (see Figure 3, next page).

But how well does the theory describe networks in the real world? It was difficult to find examples for which they could get a

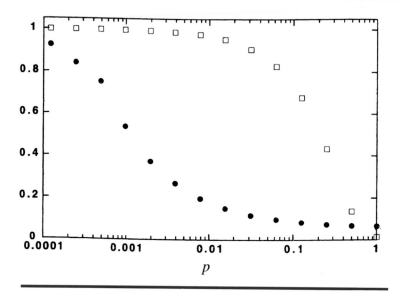

Figure 3. *Average distance between points (solid circles) and clustering coefficient (open squares) as a function of p, for a network of 1000 nodes with 10 neighbors per node. The data are scaled by their "big world" values 50 for average distance and 3/4 for clustering coefficient. (Reprinted with permission from* Small Worlds, *by Duncan J. Watts (1999), Princeton University Press.)*

complete "wiring diagram," Strogatz notes. "In the end, we could only get data for three networks," he says. There was no guarantee they would find any of the model's predictions borne out. But the theory had a high success ratio, Strogatz says: "We were three for three."

One of them is the Internet Movie Database (IMDb), in which the "nodes" are movie actors, with a connection between any two who have appeared in the same movie. The IMDb underlies the popular "Kevin Bacon Game," in which the goal is to link any actor who's ever been in a movie with the star of such film classics as "Tremors" and "Footloose." (On learning of the game in 1996, Brett Tjaden and Glenn Wasson, then both graduate students in computer science at the University of Virginia, created "the Oracle of Bacon," a website that does this automatically. For example, the Oracle reveals that Lucy Lawless, the star of "Xena, Warrior Princess," fought in "The Rainbow Warrior" with Stacey Pickren, who tippled in "Barfly" with Pruitt Taylor Vince, who recently passed the time of day with Kevin Bacon in "24 Hours.")

Analyzing the IMDb network, which then had an average of $k = 61$ connections per actor for $N = 225{,}226$ actors (the IMDb has now more than doubled in size), Watts and Strogatz found a clustering coefficient $C = 0.79$—surprisingly close to the theoretical value for a highly regular network—and an average distance between actors of 3.65, which is much closer to $\log(N)/\log(k) = 2.998$ than it is to $N/2k = 1846$.

The other two networks they analyzed were "electrical": the power grid for the western United States and the neural system of the much-studied nematode *C. elegans*. (Nematodes are a class of worms comprising 10,000 known species. The parasitic kinds, such as hookworm, infect nearly half the world's human population.) The power grid has $N = 4941$ nodes, consisting of generators, transformers, and substations, with an average of $k = 2.67$ high-voltage transmission lines per node. *C. elegans* has a mere $N = 282$ neurons (not counting 20 neurons in the worm's throat, which biologists have not yet completely mapped), with an average of $k = 14$ connections—synapses and gap junctions—for each neuron.

The power grid has a rather low clustering coefficient, $C = 0.08$. But it's still 160 times larger than what you'd expect for a random network of the same size. The average distance between nodes is 18.7, far below the 925 predicted for a perfectly regular network. (Actually this is an unfair comparison, since Watts and Strogatz's regular network is essentially 1-dimensional, whereas the power grid is inherently 2-dimensional. It would be fairer to compare the power grid to a hexagonal honeycomb, for which most nodes have three links (see Figure 4). The average distance between two nodes in an $n \times n$ honeycomb, which has $N = 2n(n + 2)$ nodes altogether, is approximately $2n$. For $N = 4941$, that's approximately 98.) For *C. elegans*, the clustering coefficient is 0.28 (versus 0.05 for the random model), and the average distance is 2.65 (with 2.25 calculated for the random model and 10 for the regular network).

Watts and Strogatz also studied a simple model of an epidemic on a small world network. They found, as expected, that their theoretical infection spread faster and faster as they increased the probability p. But they reported a surprise as well: "The alarming and less obvious point is how few short cuts are needed to make the world small."

Research on small world networks has proliferated in the last five years, thanks in part to the small world network of research itself. (Mathematicians have their own version of the Kevin Bacon Game, known as the Paul Erdős graph—see "Proof by Example: A Mathematician's Mathematician," *What's Happening in the Mathematical*

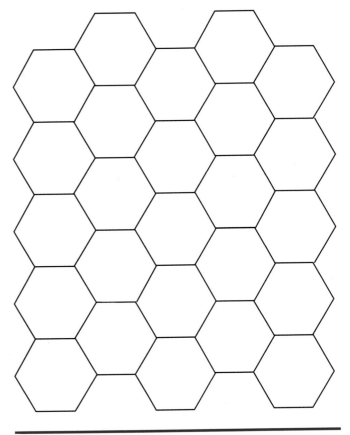

Figure 4. *A hexagonal honeycomb "power grid."*

Even when short chains exist, it may be impossible to find them.

Sciences, Volume 4.) In joint work with Mark Newman and Cris Moore at the Santa Fe Institute, Watts has derived a "mean-field" solution for a small world network similar to Watts and Strogatz's original example. For large networks, they find that adding n random links reduces average path length by a factor of $\ln(1 + n + \sqrt{n^2 + 2n})/\sqrt{n^2 + 2n}$ (where "ln" is the natural logarithm). In other words, even a single short cut ($n = 1$) saves, on average, nearly 24% ($\ln(2 + \sqrt{3})\sqrt{3} \approx 0.76$) on "travel time" between nodes in such a network!

Jon Kleinberg at Cornell University has tackled another aspect of small worlds: Even when short chains exist, it may be impossible to find them. (This might help explain, for example, why so few of Milgram's letters reached their destination.) In Kleinberg's model, each node in a two-dimensional network is connected to its four nearest neighbors and also to one distant node chosen at random, with probability that falls off with a power α of the distance, measured in the "nearest-neighbor metric." However, the nodes "know" nothing about each other's random connections, so if one node is given a "letter" addressed to another, distant node, it can only guess which one of its connections is actually closer to the target.

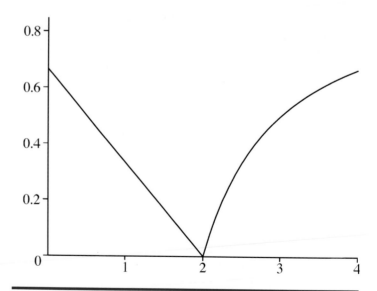

Figure 5. *Nearest-neighbor navigation. The average number of steps to deliver a message in an $n \times n$ lattice with random shortcuts based on a power-law parameter α exceeds n^β, where $\beta = (2 - \alpha)/3$ if $\alpha < 2$, and $\beta = (\alpha - 2)/(\alpha - 1)$ if $\alpha > 2$. Only for $\alpha = 2$ is the advantage of the small world network realized.*

If the power α is fixed at 2 (analogous to the inverse square power law of gravity), a "greedy" algorithm, which forwards letters to whichever node is closest in the nearest-neighbor metric, does the job in true small-world fashion: The number of steps for an $n \times n$ network grows like $(\log n)^2$. But if α is any other power, the greedy algorithm—and any other algorithm that uses only "local" knowledge of the network—takes much longer, requiring at least n^β steps, where β depends on α (see Figure 5). Roughly speaking, if $\alpha < 2$, then the first few steps usually put the letter near its target, but then the random connections tend to be so long range that they're not useful. If, on the other hand, $\alpha > 2$, then the random connections tend to be too short range to be of significant use.

While the study of small worlds has brought useful insights, the small-world phenomenon is just one aspect of the theory of networks. Of even greater interest, Strogatz thinks, is the study of dynamics in evolving networks: systems in which the nodes—and even the links between them—are dynamical systems, and which grow (or sometimes shrink) and constantly "rewire" themselves in response to their environment—or perhaps just randomly. The World Wide Web is one obvious example, but biological systems, with their intricate cycles, could be even richer.

The study of network dynamics could even be one of its own examples. Take Watts, for example. Following his Ph.D. in theoretical and applied mechanics at Cornell and postdocs at Columbia, the Santa Fe Institute, and M.I.T., Watts recently relocated to Columbia University—in the sociology department. Talk about creating a small world.

> **Research on small world networks has proliferated in the last five years.**

Goldilocks and the Three Networks
(continued)

What Goldilocks didn't know was that the spider family was not gone for the day, or even out for a walk. Instead they were debating in the kitchen. One wanted beetles for breakfast. One wanted larvae for lunch. One wanted houseflies for dinner. But when they sensed a house guest, their argument was immediately settled. On 24 legs, they tiptoed into the bedroom.

Goldilocks awoke to a trio of tickles: one on her anklebone, one on her wristbone, one on her cheekbone. It was then she discovered something else she hadn't counted on: Spider webs are sticky—and strong.

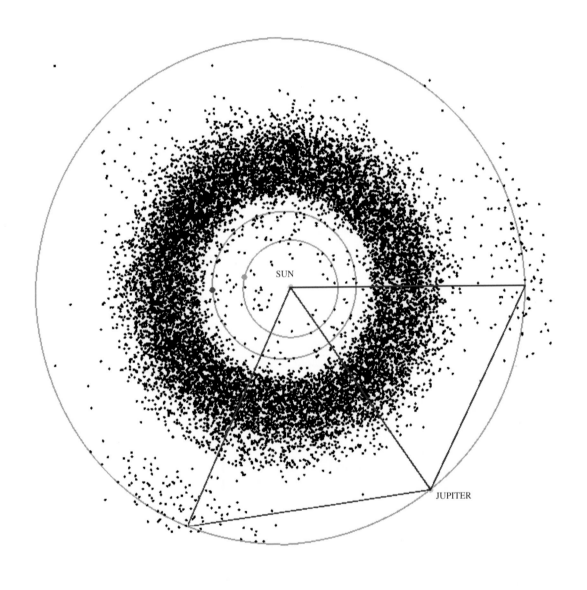

N-body Ballet. *Two clusters of asteroids—the Trojans—join Jupiter and the Sun in orbits that approximate Lagrange's solution to the 3-body problem. The slightly elliptical orbits of Earth, Mars, and Jupiter are shown. (Figure courtesy of the Minor Planet Center at the Smithsonian Astrophysical Observatory.)*

A Celestial *Pas de Trois*

In celestial mechanics, it takes two to tango. A single star in otherwise empty space traces a simple straight line—from its own point of view, it just sits there. A dull dance. But as soon as a partner appears the ballet begins. Two stars (or a star and a planet) perform a gravitational *pas de deux*. Each sweeps out a conic section, the most notable being the ellipse, first made famous as an orbital option by the mathematician–astronomer Johannes Kepler.

But what happens when there are three bodies, or four, or N? Until recently, researchers knew only a few simple dance steps with more than two bodies. That's all changed, says Richard Montgomery, a dynamical systems expert at the University of California at Santa Cruz. A combination of mathematical analysis and computer simulation has revealed thousands of new "choreographies" in which idealized, pointlike particles follow precise, periodic orbits.

Montgomery and Alain Chenciner at the University of Paris began the new dance craze in 1999, with the analysis of a new orbit in which three equal masses pursue one another along an endless,

> **Until recently, researchers knew only a few simple dance steps with more than two bodies.**

Richard Montgomery. *(Photo courtesy of Alessandra Alvares.)*

figure-eight trajectory (see Figure 1). Their results were guided in part by computer simulations by Carles Simó at the University of Barcelona, who has since found similar figure-eight orbits with up to hundreds of masses. Simó has also computed millions of new 3-body solutions with other shapes. Other researchers have joined in, including Joseph Gerver at Rutgers University in Camden, New Jersey, who found a 4-body orbit shaped like a double figure eight (see Figure 2).

The new orbits have been dubbed "choreographies." Viewed in computer animations, some of the orbits suggest the patterns followed by marching bands, with bodies barely missing one another as they whiz along. The plethora of new solutions to the equations of gravitational motion provide new tools for studying celestial mechanics. It's even possible, researchers say, that somewhere in the universe—maybe even in our own Milky Way—three or more stars have been locked for billions of years in one of these newfound ballets.

What exactly has been found, and what's so new about it? After all, the equations of gravitational motion have been known since 1687, when Newton finally wrote his *Principia mathematica*: Point particles exert forces on each other proportional to the product of their masses and inversely proportional to the square of the distance between them, and each mass accelerates in response to the forces acting on it. And it's known—isn't it?—that once you kickstart a bunch of particles, their fate is determined for all time. To determine, for example, where the planets will be in a billion years, just figure out where they are now, what directions they're going, and with what speeds (you also need to know how heavy each one is), and then solve Newton's equations of motion. End of story.

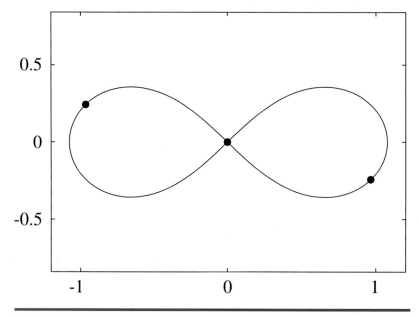

Figure 1. *Montgomery and Chenciner's figure-eight orbit. As the three bodies proceed along the loop, they occasionally line up with one at the center. (Figure courtesy of Richard Montgomery,* Notices of the AMS, *May 2001.)*

Not so fast. There are just a few hitches. One problem is that the *true* equations of gravitational motion have been known for less than 100, not 300, years. They stem from Einstein's General Theory of Relativity, and they're a heck of a lot harder even to state, let alone solve, than the classical equations. But this is just a quibble. The classical theory of celestial mechanics, using Newton's inverse-square law, is an accurate approximation in many settings. Besides, one can still ask the mathematical question: What would happen *if* Newton's version were exactly correct?

A second problem concerns uncertainties. If you don't know exactly where the planets are now, exactly where they're headed and how fast, and exactly how massive they are, you're not likely to be able to figure out even approximately where they'll be in the distant future. A third, closely related problem is that, in general, the equations of motion for celestial mechanics cannot be solved even if the initial conditions *are* known exactly. This is one reason periodic orbits are important: They provide a toehold in an otherwise intractable problem.

Collisions present a fourth problem. When particles collide, the equations of motion are momentarily meaningless: An infinite force acts in an indeterminate direction on bodies traveling at infinite velocity. If only two bodies are involved the problem is fairly easy to fix. In essence, you can assume that the system continues as if there had been a near miss. But triple collisions are a mathematical mess—researchers have no satisfactory theory that handles all possible three-fold collisions. Worse yet, there's no way to guarantee an N-body configuration isn't fated for collision, unless the bodies are clearly flying apart—or if their motion is provably periodic.

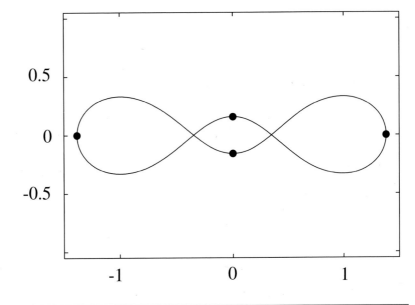

Figure 2. *Gerver's 4-body orbit. The equal masses are positioned at the vertices of a shape-shifting parallelogram, which varies from a rhombus (shown here) to a rectangle. (Figure courtesy of Richard Montgomery,* Notices of the AMS, *May 2001.)*

The only part of celestial mechanics that is completely understood is the motion of two bodies.

The only part of celestial mechanics that is completely understood is the motion of two bodies. As Newton deduced, an inverse square law of gravitational attraction, combined with the law "force equals mass times acceleration," implies Kepler's observation that planetary orbits are conic sections. In particular, given the right initial conditions, two point particles of mass m and M will follow elliptical orbits about their joint center of mass. (If one particle is much heavier than the other, then the center of mass is extremely close to the heavy particle.) Two particles can collide only if they start out with both headed directly toward their center of mass. The collision trajectories can be thought of as extremely "eccentric" (i.e., long and skinny) ellipses. In effect, the particles collide elastically: Each one rebounds and then retraces its inward path.

A crucial property of 2-body motion is that elliptical orbits are stable: Small changes to either particle's mass or trajectory causes correspondingly small changes to the orbits. Stability disappears, however, when a third dancer appears. Most N-body orbits are like a house of cards: In theory they can stay fixed forever, but the slightest perturbation wrecks the balance.

The newfound figure-eight orbit is especially interesting: Simó's numerical experiments show it is quite stable. All previously known 3-or-more-body orbits were unstable unless one of the masses was much heavier than the others—and often not even then.

For the case of three bodies, only three periodic orbits were known before the recent discoveries. One of these, known as "tight binaries," was analyzed in 1878 by the mathematical astronomer George Hill. It essentially models the Sun, Earth, and Moon: Two nearby bodies orbit each other while simultaneously orbiting a third, distant mass. The system is stable if the angular momentum is large enough to keep things from spiraling into each other.

The other two solutions date back to Euler and Lagrange, in 1767 and 1772, respectively. Both solutions are instances of what are called "central configurations," which have unsolved problems of their own (see Box, "A Delicate Balance," page 74). In Euler's solution the three bodies stay in a straight line. The simplest example has two equal masses circling a third. In Lagrange's solution the bodies stay at the vertices of an equilateral triangle. If the masses happen to be equal, the triangle simply rotates as the bodies trace out a circle. With three unequal masses, each body traverses its own ellipse (or circle), and the triangle changes size—but not shape—during the orbit (see Figure 3).

Lagrange's solution is stable when one body greatly outweighs the other two. For circular orbits, the pertinent inequality is $27(m_1 m_2 + m_2 m_3 + m_3 m_1) < (m_1 + m_2 + m_3)^2$. Indeed, our own solar system sports such an orbit: The Sun, Jupiter, and a cloud of asteroids known as the Trojans are locked in one of these equilateral orbits (see Figure "N-body Ballet," page 68). The Euler solution, on the other hand, is always unstable.

The new figure-eight orbit was actually first discovered numerically, in 1993, by Cris Moore at the Santa Fe Institute. Montgomery and Chenciner rediscovered it independently in 1999, while pursuing an idea suggested by Richard Moeckel at the University of Minnesota. Moeckel's idea was to study orbits in terms of trajectories in "shape space." For three bodies, the shape space is a sphere: Each point of the sphere corresponds to the shape of a triangle. (The north and south poles correspond to the clockwise and counterclockwise orientation of the equilateral triangle; points on the equator correspond to "degenerate" triangles, whose vertices lie on a common line.) As the bodies move, the shape of the triangle they describe changes (unless it's a Lagrange- or Euler-type orbit), producing a kind of "shadow" trajectory on the sphere. For periodic orbits, the corresponding trajectory in shape space is also periodic.

The converse is false: A nice closed loop in shape space need not be the shadow of *any* Newtonian motion, much less a periodic orbit. But orbitally meaningful loops can be identified using techniques from the calculus of variation. This theory, which dates back to Euler, takes ordinary calculus a giant step further. In ordinary calculus, the variables are real numbers; in the calculus of variations, the variables are curves. The calculus of variations problem is to find a loop that minimizes an integral known as the "action." The existence of such loops is not in doubt—the tricky bit is showing that the minimizer doesn't include a collision.

Montgomery began by looking for orbits whose "shadows" in shape space were certain simple kinds of loops. Chenciner, asked to check some of Montgomery's results, found a flaw in one of the arguments involving unequal masses. Montgomery's equal-mass example survived, but in checking the math Chenciner realized that all three masses were all traveling on the *same* curve: the now-famous figure eight. (Amusingly, Montgomery's flawed preprint was already titled "Figure 8s with three bodies," but the figure eights he had in mind were loops in shape space!)

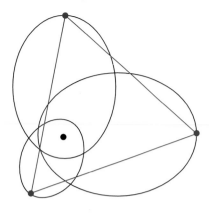

Figure 3. *Equilateral triangles are the basis of Lagrange's solution to the 3-body problem. The masses in this example are in ratio 1:2:4 (with the largest mass having the smallest orbit).*

Seeing that all three bodies trace out the same curve enabled Chenciner and Montgomery to simplify their proof that no collisions occur.

Seeing that all three bodies trace out the same curve enabled Chenciner and Montgomery to simplify their proof that no collisions occur. Simó's simulations have given additional insight, although his computational observation of the orbit's stability has not yet been backed up with mathematical proof.

Using the same techniques that nailed the 3-body figure eight, Simó has computed myriad other examples with up to hundreds of equal masses on a single closed curve. As convincing as the simulations are, the theory has some catching up to do: So far only the Montgomery–Chenciner solution is rigorously known to exist. Chenciner and Montgomery, along with Gerver and Simó, have begun a systematic analysis of "simple choreographies" for N bodies. ("Simple" means that all bodies adhere to one curve. But theorists envision other possibilities, in which some bodies travel on one curve while others orbit on another.) The four researchers have been able to prove the existence of solutions, but only when they replace Newton's gravitational force with one that falls off with the *cube* (or any higher power) of the distance between bodies. Some of these solutions seem to persist, at least numerically, in the Newtonian case, but others do not. Like all good discoveries, the figure-eight orbit has opened a universe of new questions.

A Delicate Balance

Both the Lagrange and Euler solutions are "central configurations." These are periodic, N-body orbits with a high degree of symmetry, in which the force experienced by each particle is always directed toward the system's center of mass and the particles move in such a way that their relative positions remain geometrically similar. (In Lagrange's solution, an equilateral triangle shrinks and grows as it rotates.) When the *size* of the shape stays constant—that is, when the particles simply travel in circles around their center of mass—the configuration is called a *relative equilibrium*.

The Lagrange and Euler solutions cover all possible relative equilibria for $N = 3$. For larger N, there are myriad more solutions. Or are there?

In 1941, Aurel Wintner conjectured that, for each set of N masses, only a finite number of relative equilibria exist. For $N = 3$, for example, the number is known to be 5: three Euler-type solutions (one for each mass in the middle), and two Lagrange-type solutions (depending on whether the equilateral triangle is oriented clockwise or counter-clockwise.) Wintner's conjecture is still open. Only one special case has been solved: In 1995, Alain Albouy at the University of Paris showed that for four equal masses there are exactly 50 relative equilibria.

Most researchers in celestial mechanics believe Wintner's conjecture. But Gareth Roberts at the College of the Holy Cross in Worcester, Massachusetts, has thrown a little sand into the machine. He has found five masses that admit infinitely many relative equilibria! Roberts starts with four particles (of equal mass) at the vertices of a square, with a fifth particle at the center. He then shows that, if the fifth particle has just the right mass, it's possible to squeeze the square into an arbitrarily thin rhombus and still have a relative equilibrium—in other words, Roberts's five particles have a whole continuum of relative equilibria.

This would kill Wintner's conjecture, but for one catch: To make things work Roberts has to assign his fifth particle a *negative* mass! (In celestial mechanics, masses are tacitly assumed to be positive.) Nevertheless, Roberts's "counterexample" suggests that Wintner's conjecture is far from obvious. (His result may also have implications for theories such as electromagnetism and fluid flow, in which forces can be repulsive as well as attractive.)

Roberts has also been studying the stability of relative equilibria. In the early 1990s, Richard Moeckel at the University of Minnesota conjectured that, in order to be stable, any relative equilibrium must have one "dominant" mass. Moeckel proved this for the special case in which $N-1$ equal masses are equally spaced around a circle with one more mass at its center: For N less than 8, he showed, *no* orbit is stable, no matter how heavy the central mass. For 8 or more masses, stability results if the central mass is sufficiently large.

Roberts has refined Moeckel's result by establishing what it takes to be "sufficiently large": For the central particle to hold onto $N-1$ bodies of mass 1, its mass must exceed a certain value c_N, which is roughly equal to $.435(N-1)^3$. (As an amusing application, this result limits the number of one-ton satellites that could stably orbit the Earth. Since our planet weighs about 6×10^{21} metric tons, and the cube root of $6/.435$ is around 2.4, we'd better not launch more than about 24 million such satellites.) And in another step toward proving Moeckel's conjecture, Roberts has shown that *no* relative equilibrium with N equal masses is ever stable if N exceeds 24,305. What this suggests is how difficult these problems are to analyze—that final number should really be only 2.

In celestial mechanics, masses are tacitly assumed to be positive.

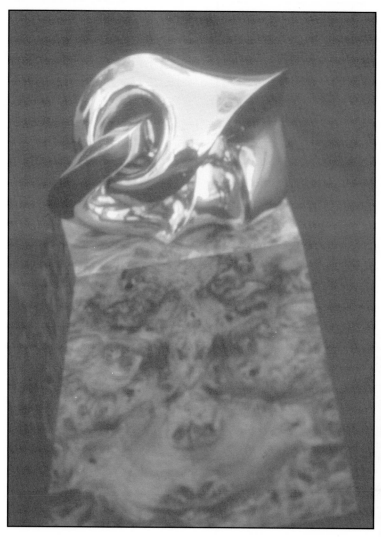

Bronze, Granite, Clay. *"Figure Eight Knot Complement vii/CMI" is mathematician–sculptor Helaman Ferguson's icon for the Clay Mathematics Institute. The large piece (bottom), done in Inner Mongolian Black Granite, is located at the CMI in Cambridge, Massachusetts. Smaller bronzer versions (top) are given as the annual Clay Research Award. (Photos courtesy of Helaman Ferguson.)*

Think and Grow Rich

It's a common question encountered by students of mathematics: If you're so smart, why aren't you rich? In fact, math majors usually *do* place near the top of the list of entry-level salaries for college graduates. But, as a quick route to easy street, immersing oneself in abstract algebra, complex functions, and topology is not the way to go.

Or is it?

The Clay Mathematics Institute, a non-profit foundation based in Cambridge, Massachusetts, has offered a million dollars each for solutions to seven open problems in mathematics. Announced in May 2000, the seven "Millennium Prize Problems," as the CMI calls them, are widely considered among the most important—and most difficult—unsolved problems today. Two problems come from number theory, two from topology, two from mathematical physics, and one from theoretical computer science.

The oldest Millennium Prize problem is a number-theoretic challenge known as the Riemann Hypothesis, which dates to 1859. It concerns a property of what's called the Riemann zeta function, customarily defined by the formula

$$\zeta(s) = 1 + 1/2^s + 1/3^s + 1/4^s + \cdots,$$

where s is a complex number. In a famous paper published in 1859, the German mathematician Bernhard Riemann used the zeta function to study the distribution of prime numbers.

Riemann's starting point was the observation, first made in 1737 by Leonhard Euler, that the zeta function can also be written as a *product* involving all primes:

$$\zeta(s) = 1/(1 - 2^{-s})(1 - 3^{-s})(1 - 5^{-s}) \cdots .$$

By analyzing the zeta function, Riemann obtained amazingly precise formulas for the distribution of primes. His analysis, which was based on the then-developing theory of complex variables, was later made rigorous, and led to a complete proof of the Prime

Landon T. Clay. *(Photo courtesy of the Clay Mathematics Institute, Inc.)*

Number Theorem: The number of primes less than x is approximately $x/\ln x$ (where $\ln x$ is the natural logarithm of x).

The formulas relating the zeta function to the distribution of primes are stated in terms of the "zeroes" of the zeta function: the values s for which $\zeta(s) = 0$. The zeroes are of two types: the "trivial" zeroes at the negative even integers ($s = -2$, -4, etc.), and the "non-trivial" zeroes, all of which have real parts between 0 and 1. In essence, the non-trivial zeroes of the zeta function contain detailed information about the error term in the approximation $x/\ln x$. (The approximation that's actually used is a function known as the logarithmic integral.) Riemann thought it "very likely" that the real parts of these zeroes are all equal to 1/2—that is, that the non-trivial zeroes of the zeta function all lie on a line. "It would, of course, be desirable to have a rigorous proof," he added.

Indeed it would. The Riemann Hypothesis, as the conjecture came to be called, has baffled mathematicians for the last century and a half. If true—and no one seriously doubts that it is—the Riemann Hypothesis has profound implications for number theory. And beyond mathematics as well: In the last decade, mathematical physicists have found echoes of the Riemann Hypothesis in the theory of quantum chaos (see "A Prime Case of Chaos," *What's Happening in the Mathematical Sciences*, Volume 4).

The other number-theoretic prize problem is known as the Birch–Swinnerton-Dyer conjecture. In the 1960s, British mathematicians Brian Birch and Peter Swinnerton-Dyer made an extensive computational study of the L-functions of elliptic curves (see "New Heights for Number Theory," page 2). These functions are closely related to the Riemann zeta function. In every example they studied, Birch and Swinnerton-Dyer found that the value of $L_E(s)$ at $s = 1$ predicts whether the elliptic curve E has finitely many or infinitely many rational solutions. Their conjecture is that E has infinitely many rational solutions if and only if $L_E(1) = 0$.

In fact, the conjecture says something more precise. Even when an elliptic curve has infinitely many rational solutions, they are all generated from a finite set. That is, the group of rational solutions has an abstract structure of the form $T + Z^r$, where T denotes a finite "torsion" group, Z denotes the integers (the German word for number is *Zahl*), and r is a non-negative integer called the "rank" of the curve. (This in itself is a deep theorem. It was conjectured by Henri Poincaré in 1901, and proved by L.J. Mordell in 1922.)

The Birch–Swinnerton-Dyer conjecture says that E has rank r if and only if $L_E(s)/(s-1)^r$ has a non-zero limit as s tends to 1.

The conjecture goes on to relate the value of this limit to various properties of the elliptic curve, but that's not required for the Millennium Prize. Some progress has been made in the last quarter century. In 1977, John Coates and Andrew Wiles at the University of Cambridge proved that, for a special class of elliptic curves with a property known as complex multiplication, if $L_E(s)$ has a nonzero limit at $s = 1$, then the curve's rank is 0, as the conjecture predicts. In 1983, Benedict Gross at Brown University and Don Zagier at the University of Maryland showed that, for "modular" elliptic curves—that is, for all elliptic curves with an associated modular form—if $L_E(s)/(s-1)$ has a nonzero limit at $s = 1$, then the rank is at least 1. In 1990, Victor Kolyvagin at the Steklov Institute in Moscow strengthened these results. He showed that the Coates–Wiles theorem holds for modular elliptic curves (a class that was already known to include curves with complex multiplication). Kolyvagin also showed that in the Gross–Zagier theorem, the rank is *exactly* 1.

Now that all elliptic curves are known to be modular—thanks to the Taniyama–Shimura conjecture having been proved—Kolyvagin's result holds across the board. But it falls short of the Birch–Swinnerton-Dyer conjecture, because it's silent about curves for which $L_E(s)/(s-1)^r$ is nonzero at $s = 1$ for r greater than 1. Conceivably, some elliptic curve lurks in the back alleys of number theory with $L_E(s)/(s-1)^2$ nonzero at $s = 1$, but with rank something other than 2—perhaps even 1 or 0.

The more famous of the two topology problems is known as the Poincaré conjecture. Loosely speaking, it asserts that a topologically complicated three-dimensional space (or "manifold") cannot masquerade as something simple. Stated technically, the conjecture asserts that a compact 3-d manifold has a trivial fundamental group if and only if the manifold is homeomorphic to the 3-sphere. A similar result is known to hold for 2-d manifolds.

"Homeomorphic" is a fancy way of saying that two things are essentially the same. In the rubber-sheet metaphor of topology, it means that a balloon is a balloon is a balloon—until you puncture, rip, or tear it. A manifold

Even when an elliptic curve has infinitely many rational solutions, they are all generated from a finite set.

Arthur Jaffe. *CMI President. (Photo courtesy of Bachrach Photographers.)*

is something that, locally at least, looks like ordinary euclidean space, be it the 2-d plane, 3-d space, or something of higher dimension. The easiest manifolds to picture are two-dimensional; these are simply curved surfaces. The "compact" ones, which include the sphere, the torus, and a host of more complicated, pretzel-like surfaces (see Figure 1), are all bounded.

The fundamental group is an algebraic structure associated with closed curves in a manifold. In the 2-d case, picture a loop of wire that is confined to the surface, but may otherwise be moved around, stretched or shrunk. For the 2-d sphere, it's easy to see—though not quite so easy to prove—that any such loop, no matter how complicated, can be shrunk down to a single point. For the torus, by contrast, there are loops that cannot be shrunk to a point, namely the ones that go all the way around, or through, the "hole" in the torus. The upshot is that the fundamental group of the 2-sphere is "trivial," whereas the fundamental group for the torus (as well as all the other, more complicated surfaces) is non-trivial.

Much the same happens with 3-d manifolds. The fundamental group of the 3-sphere is trivial, but the fundamental group of the 3-d torus is not. In 1904, Henri Poincaré conjectured that all other 3-d manifolds also have non-trivial fundamental groups. This is certainly true for all *known* 3-d manifolds. Unfortunately, the classification of manifolds, which is extremely simple in two dimensions, is itself still conjectural in 3-d. Indeed, the best way to prove the Poincaré conjecture—"best" because it would prove so much more—would be to prove the Thurston Geometrization Conjecture (see "Topology Makes the World Go Round," page 38).

Curiously, higher-dimensional analogues of the Poincaré conjecture have already been proved, even though the classification theory in higher dimensions is extremely primitive. The analogues say that an n-dimensional manifold has trivial

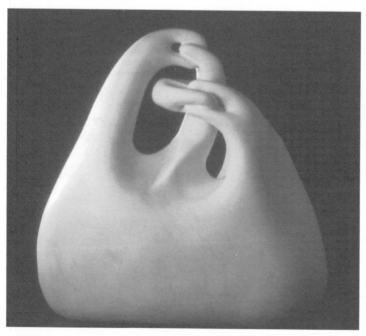

Figure 1. *In Whaledream II, mathematician–sculptor Helaman Ferguson portrays the quiet complexity of topological space. (Photo courtesy of Helaman Ferguson.)*

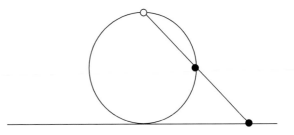

Sphere Basics

What is the 3-sphere? It's best approached through lower dimensions. The 1-sphere is simply the ordinary circle. The theory of compact 1-d manifolds is extremely simple: The 1-sphere is the only one. (Incidentally, the fundamental group of the 1-sphere is not trivial: The circle *itself* is a loop, and there's no way you can contract it to a point.) The 2-sphere is the sphere of ordinary speech. Like the circle, it can be defined algebraically, by the equation $x^2 + y^2 + z^2 = 1$. Topologists call this an "embedding" of the 2-sphere in R^3. A nice property of 2-d manifolds is that every (orientable) 2-d manifold can be embedded R^3.

The 3-sphere is most easily defined algebraically, as the solution set in R^4 of the equation $x^2 + y^2 + z^2 + \omega^2 = 1$. Unfortunately, objects in 4-dimensional space are hard to picture. However, there's another way to understand the 3-sphere that leads to a fairly good grasp of it. We start again with the 1- and 2-spheres.

Another view of the circle is as the entire real line with one extra point added to close the loop. The correspondence comes courtesy of a simple map called the stereographic projection (see Figure 2). Similarly, the 2-sphere is the entire infinite plane with one extra point (sometimes called the north pole). In other words, the 1-sphere is much like a line, and the 2-sphere is much like a plane—with the one difference that they've been "compactified" by the addition of one extra "point at infinity." Again, the 3-sphere is similar: It's basically just euclidean 3-d space with one slightly mysterious extra point glued on.

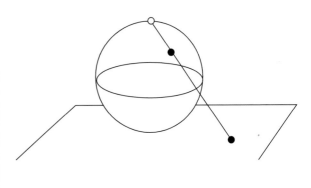

Figure 2. *Stereographic projections equate the circle and the sphere with the line and the plane, respectively, each with an extra "point at infinity."*

fundamental group if and only if it's homeomorphic to the n-sphere. For dimensions 5 and higher, this was first proved in 1960 by Stephen Smale at the University of California at Berkeley. The 4-d case resisted until 1982, when Michael Freedman at the University of California at San Diego knocked it off.

One reason the Poincaré conjecture is so hard in 3-d has to do with the nature of the surface that's swept out when a loop is contracted to a point. The surface is topologically a disk—that is, it looks like the interior of a circle—but it can wind up embedded with all kinds of kinks and self-intersections. This can't happen for loops on the 2-sphere. But loops in 3-d are often knotted, and that makes for a mess. Even in 4-d it may be impossible to avoid self-intersections when contracting to a point. But in 5-d and higher, there's enough "room" to maneuver, so that the disk can always be swept out in a nice way.

The other million-dollar topology problem is the Hodge conjecture, named for William Hodge, who first proposed it in 1950. The Hodge conjecture concerns the analysis of high-dimensional manifolds defined by systems of algebraic equations. It says, very roughly, that everything you always wanted to know about algebraically defined manifolds (but were afraid to ask) is to be found in the theory of calculus.

An algebraically defined manifold is a space that's defined by a system of polynomial equations. As a rule, one equation in n variables defines a "hypersurface" of dimension $n - 1$. For example, the equation $x^2 + y^2 + z^2 = 1$ in 3 variables defines the 2-dimensional sphere. For a system of, say, k equations, the intersection of the corresponding hypersurfaces is, in general, of dimension $n - k$.

The high-dimensional analogues are impossible to picture, especially if you let each variable be a complex number, but the basic mathematical principles carry over from lower-dimensional settings. In particular, it's possible to do calculus on these manifolds. The formalism goes by the name "cohomology" theory. At the heart of this theory is a vector space of differential forms known as the Hodge space. (In calculus, a differential $f\,dx$ is "what gets integrated" in the integral $\int_a^b f(x)dx$. Differential forms are higher-dimensional generalizations.) The Hodge conjecture asserts that, for each algebraically defined manifold, this vector space is spanned by differential forms associated with algebraically defined submanifolds.

The precise statement of the conjecture is highly technical and the concepts involved are extremely subtle. In fact, Hodge himself first stated the conjecture incorrectly! (He also proposed a generalization that turned out to be incorrect. The French mathematician Alexandre Grothendieck corrected the latter in a 1969 paper with an eye-catching title: "Hodge's general conjecture is false for trivial reasons.") Such missteps are common when fields are first developing. It takes time for researchers to determine which analogies are the most appropriate.

By now mathematicians understand the theory well enough to be confident the Hodge conjecture is properly stated, and they expect that it's correct. Still, the conjecture could be false for "trivial" reasons that have escaped notice for the last 50 years. The CMI is willing to award a prize for a counterexample to any of its problems if the counterexample radically alters the nature of the theory, but reserves the right to reformulate the problem if, as happened with Grothendieck, the counterexample reveals only a minor flaw. (Due to the substantial sums involved, the CMI has written its rules very carefully.)

The two Millennium problems in mathematical physics are concerned with our mathematical understanding of two familiar features of the real world: the fact that water (or any other fluid, including the air we breathe) flows in a smooth (if sometimes turbulent) manner, and the fact that matter has mass—i.e., the fact that there's something for force to accelerate!

The first problem refers to the existence and smoothness of solutions to the Navier–Stokes equation (which actually consists of two equations). The Navier–Stokes equation essentially applies Newton's famous formula (force equals mass times acceleration) to fluids: It describes the way that fluid flow changes in space and time.

The equation is extremely difficult to solve in general. Exact solutions are possible in only a few settings. (For example, if the fluid starts out perfectly still, then, according to the Navier–Stokes equation, it remains so for all time. But as solutions go, that one's about as dull as ditchwater!) For the most part, fluid-flow scientists, which include a range of researchers from pure mathematics to aerodynamics, are content to use numerical algorithms that produce approximate solutions. But they would love to know more about the exact solutions. As a first step, they would simply like to know that exact solutions really do exist, no matter what the initial conditions.

And there's the hang-up: It's *not* known that the Navier–Stokes equation always has a solution. It's conceivable that a fluid could begin flowing in such a way that it eventually develops a "singularity"—a point or points where the flow is no longer continuous,

It's not known that the Navier–Stokes equation always has a solution.

The 3-d obstacle is turbulence.

such as a vortex around which the fluid spins with greater and greater velocity the closer in you look. The Millennium problem is to prove that this can't happen: If the initial conditions are smooth, then the flow remains smooth for all time.

In fact, the Millennium problem could go either way: You can win the prize by finding a smooth set of initial conditions for which the flow *doesn't* stay smooth. In two dimensions, the analogous problem was solved in the 1960's, by Olga Ladyzhenskaya at the Steklov Institute in St. Petersburg, Russia, who showed that smooth conditions necessarily spawn smooth flows. And partial results for the 3-d case are known: All smooth flows stay smooth for at least a short amount of time, and if a flow is slow enough to start with, then it stays smooth forever.

The 3-d obstacle is turbulence. To keep track of a turbulent fluid, you need to make finer and finer observations of it: The fluid continually generates features at smaller and smaller scales. Turbulence does not exist in two dimensions, but it plagues the 3-d theory. Very roughly speaking, the question of existence and smoothness for the Navier–Stokes equation amounts to asking how quickly turbulent flow reaches these smaller and smaller scales. If, say, you have to double the magnification once per second, then the flow will always be seen as smooth—after n seconds, you just have to look 2^n times closer. But if you have double the magnification first after one second, again just a half second later, yet again just a quarter second later, and so forth, then you're going to have a mess in two seconds.

While the Navier–Stokes equation concerns classical mechanics, the other physics problem on the CMI's hit list comes from quantum theory. The problem is to show that the framework mathematical physicists use to study quantum fields—an approach known as Yang–Mills theory—really does describe the subatomic world of quarks, gluons, and the rest of the particle zoo. In particular, the theory should predict a "mass gap" in the energy spectrum. Loosely speaking, this means that in a theory of, say, quarks, the energy of empty space is 0, but as soon as even one particle appears, the energy is at least some minimum value, E. Assuming this, you can assign the particle the mass m via Einstein's formula $E = mc^2$.

The current version of Yang–Mills theory, known as quantum chromodynamics (QCD), almost certainly predicts a mass gap, along with other desirable properties such as asymptotic freedom and quark confinement. (The former means that at shorter and shorter distances, the quantum behavior of the field is better and better approximated by the classical theory; the latter explains why we live in a sea of protons and pions and such, rather than an ocean of quarks.) Computational studies of QCD indicate the theory works, and experiments to date confirm its predictions. But rigorous proof is lacking.

The official problem goes beyond chromodynamics. QCD is only one Yang–Mills theory, based on a "gauge" group of symmetries known as $SU(3)$. (Roughly speaking, $SU(3)$ is the group of rotations in 3-d complex space.) The Millennium problem is to establish a Yang–Mills theory for *all* gauge groups, and to show that each one has a mass gap. A tall order indeed, but theoretical physicists consider it essential to a proper understanding of quantum fields.

The final Millennium problem, called the "P versus NP" problem, goes to the heart of computer science. At issue is whether many problems that are now out of reach computationally will always remain so. In other words, are there computational problems that computers will *never* handle?

Technically, both P and NP are classes of "decision" problems, which seek a simple Yes/No answer. For example: Does 1,002,011 have any divisors less than 1000? Many optimization problems, such as finding the shortest route in the travelling salesman problem, can be reduced to a set of such Yes/No questions.

For problems in class P, the work required to solve a representative problem grows no faster than a *polynomial* in the "size" of the problem. To decide, for example, whether two N-digit numbers have a common factor requires a computation—the Euclidean algorithm—that grows like N^2. Such problems are relatively "easy," because the extra time needed to solve a slightly bigger instance is a tiny fraction of the time spent on the smaller version. If it takes, say, 1 second to find the greatest common factor of two 1000-digit numbers, then it should take only 2 more milli-seconds to handle two 1001-digit numbers.

The class NP—the moniker means *non-deterministic polynomial time*—is trickier to describe. Loosely speaking, NP consists of problems for which proposed solutions are easy to *check*. Factorability is a good example: The answer to the question about

> The "P versus NP" problem goes to the heart of computer science. At issue is whether many problems that are now out of reach computationally will always remain so.

1,002,011 is Yes, as you can easily check on a pocket calculator by dividing it by 733. "Easy" means that the proposed solution can be verified with a computation that grows polynomially in the size of the problem. But the proposed solution can come from anywhere: an exhaustive search, a lucky guess, or some sort of insider knowledge. That's what makes them non-deterministic.

More precisely, NP consists of Yes/No problems that are easy to check *when the answer is Yes*. The flip side is called co-NP. Asking whether a number is prime, for example, is co-NP, since it's easy to demonstrate a factor if you happen to know one. (Actually primality/factorability is in *both* NP and co-NP. Number theorists have developed a theory of "prime certificates," which enables one to verify that a large number is prime with a polynomial amount of computation.)

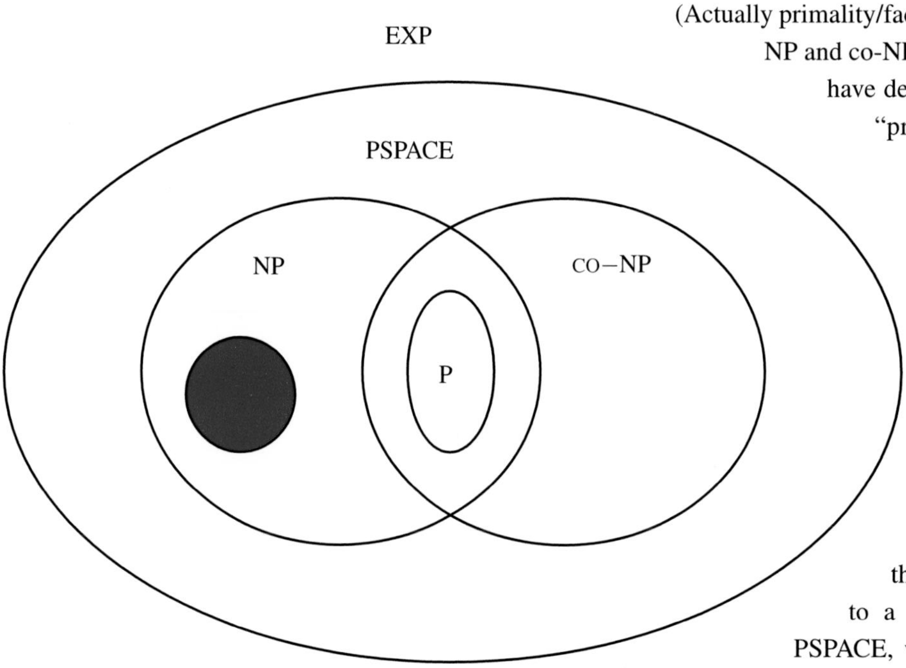

P is a subset of both NP and co-NP. All three classes belong to a larger class called PSPACE, which belongs to a yet larger class of problems called EXP (see Figure 3). EXP consists of Yes/No problems that can take an exponential amount of time to solve, possibly because they require intermediate results that have exponentially many digits (for example, does the decimal expansion of 2^{2^n} contain an even number of 1's?). PSPACE problems demand less extensive bookkeeping. You can think of them as problems that can be worked out in an exponential amount of time on a polynomial-sized—but *erasable*—blackboard.

Computer scientists believe that P is *strictly* smaller than NP, that NP is strictly smaller than PSPACE, and that PSPACE is strictly smaller than EXP. All that's now known for sure is that P is strictly smaller than EXP—there are indeed problems that cannot

Figure 3. *The hierarchy of complexity theory. The colored circle indicates the class of NP-complete problems.*

be solved in polynomial time. But it may still be the case that P=NP=PSPACE, NP=PSPACE=EXP, or any other combination that doesn't involve P=EXP.

The Millennium problem focuses on P and NP because they contain most of the problems that computers face in practice. To put it jokingly, P includes the problems scientists *can* solve, while NP contains the ones they *want* to solve.

Occasionally a problem formerly only known to be in NP is shown to be in P. The problem of linear programming enjoyed such a demotion in 1979, when Leonid Khachian found a polynomial-time algorithm for it. (Khachian's method was dramatically improved by Narendra Karmarkar at Bell Labs in 1984.) But some NP problems belong to P only if *all* NP problems belong to P. These problems are called NP-complete. The concept was introduced in 1971 by Stephen Cook at the University of Toronto, who gave the first such example. Thousands of problems are now known to be NP-complete (see "Ising on the Cake," page 88). In essence, each one contains all the difficulties of anything in NP. So if you want to find a problem in NP can't be solved in polynomial time, you may as well focus on the ones that are NP-complete.

> **Computer scientists believe that P is *strictly* smaller than NP, that NP is strictly smaller than PSPACE, and PSPACE is strictly smaller than EXP.**

Alternatively, you might *find* a polynomial-time algorithm for an NP-complete problem. If you do, then you have, in effect, found polynomial-time algorithms for *all* NP problems, and the Clay Mathematics Institute will make you a millionaire. But that million will pale beside what you'll gain by cracking all the cryptographic systems based on the difficulty of NP problems.

The CMI, whose institutional mission is to "further the beauty, power, and universality of mathematical thought," hopes that the prize offer will not only attract new efforts to solve these seven important problems, but recruit more newcomers to mathematics. The intellectual rewards are enormous. The pay's not bad, either.

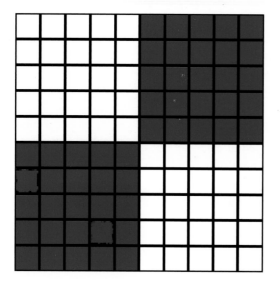

Spin City. *To each square on a "randomly" colored grid, assign a value +1 or −1. Multiply the values of adjacent squares, with a "coupling coefficient" −1 if the squares are both white (and +1 otherwise). Add up all the resulting products. In some cases it's easy to find an assignment that gives the largest possible total. In other cases, it's not so easy. (Bottom design courtesy of Loren Larson.)*

Ising on the Cake

For decades, the gold standard in statistical mechanics has been a mathematical problem known as the Ising model. Introduced in the 1920s by German physicist Ernst Ising (who later taught at Bradley University in Peoria, Illinois), the Ising model is a theoretical framework for the study of phase transitions: the abrupt changes of state that occur, for instance, when ice melts or cooling iron becomes magnetic. While they've learned much from approximate solutions and computer simulations, physicists have long sought an exact mathematical solution, which would provide much more information about phase transitions.

Well, as they say in New York, fuhgeddaboudit.

Sorin Istrail, a theoretical computer scientist at Sandia National Laboratory (now at Celera Genomics in Rockville, Maryland), has shown that the Ising model—or, more precisely, a version known as the 3-d spin glass model—belongs to a class of recalcitrant calculations known as NP-complete problems. His proof extends earlier work of Francisco Baharona of the University of Chile. In effect, an exact solution to the Ising model would provide the key to efficient solutions of thousands of other computational problems, ranging from the famous problem of the traveling salesman to the task of factoring large numbers.

Although the existence of efficient algorithms for any of these problems has yet to be ruled out, theorists are fairly certain that NP-complete problems really are as hard as they seem to be. Proving a problem is NP-complete is regarded as strong evidence that it is effectively unsolvable by computers. Indeed, many cryptographic systems are based on the premise that these problems are truly as hard as they appear (see "Tales from the Cryptosystem," *What's Happening in the Mathematical Sciences*, Volume 4).

The Ising model concerns itself with interactions between neighboring atoms (or other objects) in a regular array, which can be 1-dimensional (think beads on a string), a 2-d grid, or a 3-d lattice (see Figure 1, next page). Some simple versions of the model *can* be solved exactly. Ising himself solved the 1-d version—and found it had no phase transition. In 1944, the Norwegian chemist Lars Onsager discovered an exact formula for the 2-d model, which does possess a phase transition. But scientists have never been able to extend Ising's and Onsager's solutions to the physically realistic realm of 3 dimensions.

Sorin Istrail. *(Photo courtesy of Sorin Istrail.)*

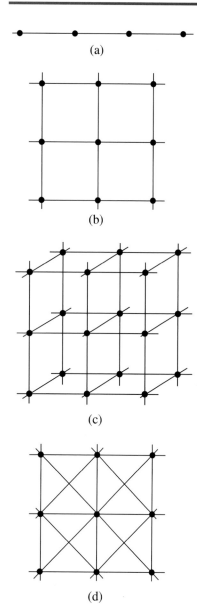

Figure 1. *The Ising model with nearest-neighbor interactions in dimensions 1, 2, and 3 (a–c), and with next-nearest-neighbor interactions in dimension 2 (d).*

"We now know why," says Istrail. "We're living in a Flatland—only planar models can be solved exactly."

Istrail's proof relies on a combinatorial interpretation of what the Ising model computes. Physically, the model tallies the number of ways a large but finite lattice can be in each of a myriad of energy levels, ranging from a highly regular "ground" state to a high-energy jumble (see Box, "To Infinity—And Beyond"). In a bar of iron, for example, each atom has a magnetic moment, or spin, that "points" either "up" or "down." Mathematically, the spin at each lattice site takes one of two values: $+1$ or -1. In the ground state the spins all point in the same direction, whereas high-energy states have lots of adjacent atoms with opposing moments. This is the original Ising model, also known as the ferromagnetic case. In the "anti-ferromagnetic" case, which was first discovered by the French physicist Louis Néel in 1932 (and for which he won a Nobel Prize in 1970), adjacent atoms prefer to point in opposite directions—that is, the nature of the interaction causes spins that point in the same direction to raise the energy and spins that point in opposite directions to lower the energy. (The read heads in modern hard disk drives rely on thin-film layers of ferromagnetic and anti-ferromagnetic materials.)

In the "spin glass" version of the Ising model, each bond connecting adjacent lattice sites is assigned its own "coupling constant," which is positive for ferromagnetic interactions and negative for anti-ferromagnetic interactions (or 0 if the sites don't interact). In the simplest case, the coupling constants take only two values: $+1$ and -1. In a typical spin glass model, the coupling constants are assigned at random.

In all versions of the Ising model, the energy of the system in any given state is the negative of the sum of all the interactions between adjacent lattice sites. Each interaction is the product of the two spins times the coupling constant. That turns the task of computing energy levels into a combinatorial problem: For a given assignment of coupling constants, find all the possible energy levels and the number of different ways to be in each. In particular, find a state of *lowest* energy.

That last task is easy enough for the ferromagnetic and anti-ferromagnetic cases. But finding a state of lowest energy for a spin glass model can be a frustrating experience. In fact, "frustration" is exactly what physicists call the effect that complicates calculations for spin glasses: In general it's impossible to assign spins that

To Infinity—And Beyond

In the Ising model—and in statistical physics generally—the probability that a system is in a state of energy E is proportional to $e^{-E/kT}$, where k is a number known as Boltzmann's constant and T is the absolute temperature. This formula implies that any given low-energy state is more likely to occur than a high-energy state. If the temperature is near absolute zero, the low-energy state is much, much more likely to occur. But because so many different states can have the same high energy, the probability of being in *some* state with high energy can be quite large.

Physicists use a function called the partition function to describe all the different possibilities and their corresponding probabilities. In essence, the partition function counts the number of different ways, $n(E)$, the system can be at each possible energy level E, and registers these counts with a relative probability $e^{-E/kT}$ for each: $P(T) = \sum n(E)e^{-E/kT}$, where the sum runs over all possible values of E.

All the pertinent information about a system, the theory says, is captured in the partition function. In particular, phase transitions occur when there is an analytic discontinuity in the *limit* of the partition function as the size of the system—i.e., the number N of lattice sites—tends to infinity. More precisely, if the function $F(T) = \lim_{N \to \infty} \log(P(T)^{1/N})$, which physicists call the "free energy," has a kink of some kind at a "critical" temperature T_c, then, by definition, there is a phase transition at that temperature. For the 2-d Ising model, the kink is a discontinuity in the second derivative of the free energy function (see Figure 2). Phase transitions of this kind are said of be of second order.

The easiest way to compute a limit and check for analytic discontinuities is if there's an explicit formula for the function in question. That's one reason researchers have been keen on computing the partition function—and one reason Istrail's NP-completeness results can be viewed as a roadblock. But complexity theory is a subtle subject. It only says that certain problems become (or seem to become) harder and harder to solve as the size of the problem grows. It doesn't say anything about what happens when a problem becomes infinitely large!

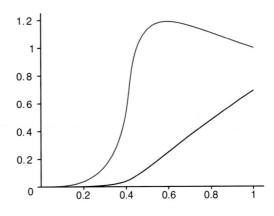

Figure 2. *The phase transition for the 2-d Ising model is hidden in the double integral*
$$\tfrac{1}{2} \int_0^1 \int_0^1 \log[(1 + t^2)^2 - 2t(1 - t^2)(\cos(2\pi x) + \cos(2\pi y))] \, dx \, dy,$$
where $t = \tanh(E/kT)$. (The free energy is this plus $\log[2/(1 - t^2)]$.) The function defined by the double integral is continuous (black curve). So is its first derivative (teal curve). But the second derivative blows up at $t = \sqrt{2} - 1 \approx 0.4142$.

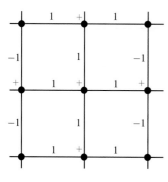

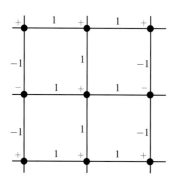

Figure 3. *Frustration. Trying to "grow" a state of lowest energy doesn't work very well. Minimizing the energy contributions of the bonds around the middle site in a simple 3×3 example leads to a total energy of -4 (top). The true minimum is -6 (bottom).*

make each interaction of low energy. If you try to "grow" a low-energy state, you almost inevitably run into assignments that raise the energy along some of the bonds, and you usually wind up with a non-optimal state of the system (see Figure 3).

When the hunt for energy levels is expressed mathematically, the formulas can be viewed as describing properties of a mathematical graph—and graph theory is one of the cornerstones of theoretical computer science. In particular, computing the ground state for the spin glass model is equivalent to finding a minimal set of weighted edges—that is, a set of edges with minimal total weight—that "cuts" the corresponding graph into two pieces. Once the cut is made, the vertices of one piece are assigned spin $+1$, while those of the other are assigned spin -1.

For planar graphs, that calculation is a relative breeze. In one dimension, the computation is trivial: Make a "cut" at each negative coupling constant (see Figure 4, top). It's a little more interesting when there is an "external magnetic field," which, graph-theoretically, means adding one more vertex that interacts with all the other lattice sites (see Figure 4, bottom). That was essentially the subject of Ising's PhD dissertation, in which he completely analyzed the 1-d ferromagnetic model with external field (and found it had no phase transition).

The 2-d case is far from trivial—Onsager's solution of the 2-d (ferromagnetic) Ising model was a tremendous theoretical breakthrough—but it's still a tractable computation. In the lingo of computational complexity theory, there is a "polynomial-time" algorithm for finding an assignment of spins that gives a state of lowest energy. More generally the partition function—that is, the number of different ways for the system to be at each energy level—can be computed in polynomial time.

In 1982, Baharona showed that the 3-d spin glass model with coupling constants $+1$, -1, and 0 is NP-complete. He did so by relating the model to a graph-theoretic problem that was already known to be NP-complete: the "Max Cut" problem (see Box, "Mission Intractable," page 94). The trick was to show that any graph, no matter how wild-looking, can be embedded in the standard cubic lattice, and the weights of its edges realized by some choice of coupling constants. This means that any algorithm for solving this special case of the spin glass model automatically gives an algorithm for solving Max Cut. Since Max Cut is NP-complete, so is the spin glass problem.

Baharona's proof also included the 2-d model with an external magnetic field. But it left other versions of the model unresolved—including the ostensibly simpler case in which the coupling constant takes on only two values. Baharona's proof also left open another model that physicists had long found inexplicably difficult: the 2-d model with next-nearest-neighbor interactions (see Figure 1d, page 90).

Istrail has now despatched all those problems: All are NP-complete. Non-planarity, rather than 3-dimensionality, turns out to be the key. Istrail's proof shows that *every* nonplanar lattice graph—including the case of a 2-d lattice with next-nearest-neighbor interactions—throws up a barrier of computational intractability.

Istrail proves first that the Max Cut problem translates into a "minimum weight cut" problem for a particular lattice graph, which he calls the "Basic Kuratowskian" (see Figure 5); this makes that problem also NP-complete. The name refers to Polish mathematician Kazimierz Kuratowski, who proved a fundamental result in graph theory known as Kuratowski's Lemma: A graph is non-planar if and only if it contains a copy of the "complete" graph with 5 vertices or the "bipartite" graph with two sets of 3 vertices (see Figure 6, next page).

Istrail next proves an analog of Kuratowski's Lemma for lattice graphs: A lattice graph is non-planar if and only if it contains the Basic Kuratowskian. The "if" part of this is rather easy; it amounts to using Kuratowski's Lemma on the Basic Kuratowskian itself. The hard part is proving the "only if."

The final step is again fairly simple. The standard cubic lattice in three dimensions and the 2-d next-nearest-neighbor lattice are easily seen to be non-planar. Combining this with Istrail's two main results shows that the general version of the spin glass model is NP-complete.

Istrail's work does not necessarily mean that all hope is lost. It might still be possible to find exact answers to specific aspects of

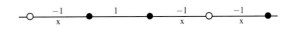
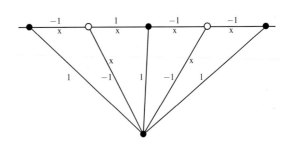

Figure 4. *Finding the best bonds to "cut" (x's) is extremely easy for a 1-d spin glass (top). It's a little trickier, but still routine, if there's an external magnetic field (bottom).*

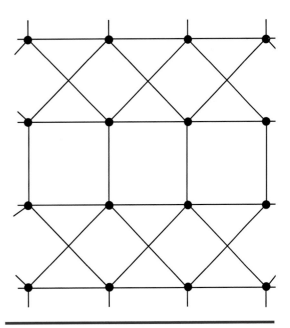

Figure 5. *The Basic Kuratowskian plays an important role in Sorin Istrail's analysis of the Ising model.*

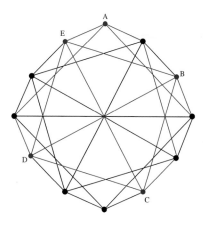

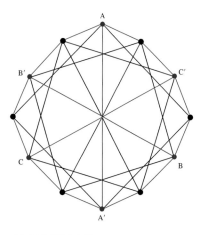

Figure 6. *According to Kuratowski's Lemma, a graph is non-planar if and only if it contains a copy of K_5, the "complete" graph with 5 vertices (top), or $K_{3,3}$, the "bipartite" graph with two sets of 3 vertices (bottom). Some graphs contain both!*

the 3-d Ising model, he notes. For example, three years before Onsager's result, Hendrick Kramers and Gregory Hugh Wannier found the exact temperature at which the phase transition occurs in 2-d ferromagnets: In the peculiar units of statistical mechanics, the crucial temperature is $\sqrt{2} - 1$. Also, the ferromagnetic case of the 3-d Ising model lies just outside of the complexity-theoretic approach: Complexity theory applies only if there are at least two kinds of interaction between pairs of nearest neighbors, but in ferromagnetism each pair interacts in the same way.

Nevertheless, "we need a paradigm shift," Istrail says. "Instead of waiting for the mathematics to advance, we have to accept this impossibility." The computational complexity of the Ising model could be just the tip of the iceberg. "There is something about this world that doesn't allow us to understand it."

Mission Intractable

It sounds so easy. Consider a graph with a bunch of vertices and edges, with weights (positive integers) assigned to the edges. You may even restrict yourself to graphs where each vertex has three edges. Your assignment (if you choose to accept it) is divide the vertices into two sets, say V_1 and V_2, so that when you add up the weights of all the edges that have one vertex in V_1 and the other in V_2, you get as large a sum as possible. One example, for a graph with 10 vertices, is shown in Figure 7. Big deal.

But what if your graph has 100 vertices instead of 10? Or 1000? In principle all you have to do is try all different ways of dividing the vertices into two sets, compute the sum of the "cut" edges for each possibility, and pick the division that comes out largest. This brute-force strategy is easy enough when there are 10 vertices: There are only 512 possibilities. But for 100 vertices the corresponding number, 2^{99}, is already far beyond what computers can handle. And for 1000 vertices—well, fuhgeddaboudit.

Is there a quicker way than brute force to solve the "Max Cut" problem? Apparently not. It belongs to the class of computational problems known as NP-complete problems. An efficient algorithm for solving any one of these problems would automatically translate into efficient algorithms for *all* problems in the class NP, which includes almost all computational problems of practical interest (see "Think and Grow Rich," page 76). Simple as it sounds, Max Cut is about as hard as a problem can get.

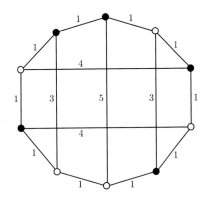

Figure 7. *Max Cut is easy for small problems. It gets harder and harder as the graph gets larger.*

Deranged Math

Why bother with fancy mathematics when computers nowadays can do it all for us? Because computers really *can't* do it all. And for what they can do, fancy math can help them do it better.

Take the case of the deranged hats. Five dapper gentlemen check their fedoras at the entrance to a nightclub. When they leave, however, the attendant returns their headgear at random. What's the probability that none of the five slightly tipsy gents winds up wearing his own hat?

The problem can be solved by brute force: Simply list all the ways the hats can be returned and then count the "derangements"—i.e., the rearrangements in which no one has his own hat. That's not a bad calculation for 5 men, because there are only $5! = 5 \times 4 \times 3 \times 2 \times 1 = 120$ possibilities. But with 10 men, the list of possibilities grows to a staggering 3 million and change (to be exact, $10! = 10 \times 9 \times 8 \times \cdots \times 2 \times 1 = 3{,}628{,}800$). And with 100 dandies and their homburgs, listing is out of the question. Brute force is practical only when the problem is small.

An approximate answer can be obtained by computer simulation: Randomly generate random returns of 5 hats and keep track of the derangements. But with only 5 hats, this is more work than it's worth: To get even a couple of decimal places of the answer requires thousands of repetitions—far more than brute-force listing. Simulations are worthwhile only for large problems.

An exact mathematical answer handles problems of *any* size. And the hatcheck problem has one. For 5 hats, the formula is

$$\tfrac{1}{2!} - \tfrac{1}{3!} + \tfrac{1}{4!} - \tfrac{1}{5!} = \tfrac{44}{120} = 0.36666\ldots.$$

For 10, it's

$$\tfrac{1}{2!} - \tfrac{1}{3!} + \cdots - \tfrac{1}{9!} + \tfrac{1}{10!} = \tfrac{1{,}334{,}961}{3{,}628{,}800} = 0.367879\ldots,$$

Moreover, elementary analysis (using first-year calculus) shows that as the number of hats gets larger and larger, the answer gets closer and closer to the number $1/e$, where $e = 2.71828\ldots$ is a number near and dear to mathematicians: It's the base of the "natural" logarithm.

Ain't math wonderful?

ORDER FORM

Ordered by

Name _____

Address _____

City _____

State/Province _____ Zip/Postal code _____

Country _____

email _____

☐ Please send me electronic notification about new AMS publications.

For orders with remittances:
(Payment must be made in U.S. currency drawn on a U.S. bank.)
American Mathematical Society
P.O. Box 5904
Boston, MA 02206-5904, USA
1-800-321-4AMS (4267)
1-401-455-4000, worldwide

For credit card orders:
American Mathematical Society
Membership and Customer Services
P.O. Box 6248
Providence, RI 02940-6248, USA
1-800-321-4AMS (4267)
1-401-455-4000, worldwide
fax 1-401-455-4046
email: cust-serv@ams.org

For Booksellers:
American Mathematical Society
Commercial Accounts Group
P.O. Box 6248
Providence, RI 02940-6248, USA
or fax purchase orders to:
1-401-455-4046
email: sales@ams.org

Shipping address (if different)

Name _____

Address _____

City _____

State/Province _____ Zip/Postal code _____

Country _____

Qty.	Code	Title	Price	Total
	HAPPENING/5	What's Happening in the Mathematical Sciences, Volume 5	$19	
	HAPPENING/4	What's Happening in the Mathematical Sciences, Volume 4	$15	
	HAPPENING/3	What's Happening in the Mathematical Sciences, Volume 3	$15	
	HAPPENING/2	What's Happening in the Mathematical Sciences, Volume 2	$10	
	HAPPENING/1	What's Happening in the Mathematical Sciences, Volume 1	$10	

	Shipping and handling	
	Residents of Canada, please include 7% GST.	
	Total	

Shipping and handling

· Orders shipped to U.S. addresses, add $3.50 for the first item and 50 cents for each additional item.

· Orders shipped surface mail to addresses outside the U.S., add $5.00 for the first item and $1.00 for each additional item. For International Air, add $13.00 for the first item and $6.00 for each additional item.

Charge by phone in the U.S. and Canada
1-800-321-4AMS (321-4267)

Payment

☐ Check or Money Order   ☐ American Express   ☐ Discover
                         ☐ MasterCard   ☐ Visa

Card No._____ Exp. Date _____

Signature _____

Please send me information about AMS membership

☐ individual membership
☐ institutional membership
☐ corporate membership
☐ institutional associate

You can also place your order through the AMS Bookstore: www.amsbookstore.org

HAP-5

What's Happening the Mathematical Sciences, Volume 5

Barry Cipra

Reviews of Previous Volumes:

This lively presentation of an amazingly wide spectrum of happenings in mathematics is impressive ... [this book] should be presented to a wide audience even outside mathematics, which could be fascinated by the ideas, concepts and beauty of the mathematical topics.

—European Mathematical Society Newsletter

An excellent source of information. Through his writing, diagrams, and sidebars, Cipra offers historical background, mathematical connections, and insight into the world of research mathematics. Throughout the book, he connects modern mathematical ideas to important applications in computer science, physics, biology, security codes, and art. He also presents an intriguing blend of historical and contemporary mathematics in each chapter. An excellent resource for high school mathematics teachers and their students.

—Mathematics Teacher

Stylish format ... largely accessible to laymen ... This publication is one of the snappier examples of a growing genre from scientific societies seeking to increase public understanding of their work and its societal value.

—Science & Government Report

The topics chosen and the lively writing fill a notorious gap—to make the ideas, concepts and beauty of mathematics more visible for the general public ... well-illustrated ... Congratulations to Barry Cipra.

—Zentralblatt MATH

Mathematicians like to point out that mathematics is universal. In spite of this, most people continue to view mathematics as either mundane (balancing a checkbook) or mysterious (cryptography). This fifth volume of the What's Happening series contradicts that view by showing that mathematics is indeed found everywhere—in science, art, history, and our everyday lives.

What's Happening in the Mathematical Sciences, Volume 5; 2002; approximately 104 pages; Softcover; ISBN 0-8218-2904-1; List $19; All AMS members $15; Order code HAPPENING/5HAP-5

Also Available ...

Volume 4	**Volume 3**	**Volume 2**	**Volume 1**
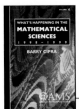	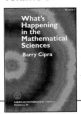		
What's Happening in the Mathematical Sciences, Volume 4; 1999; 126 pages; Softcover; ISBN 0-8218-0766-8; List $15; Order code HAPPENING/4HAP-5	**What's Happening in the Mathematical Sciences,** Volume 3; 1996; 111 pages; Softcover; ISBN 0-8218-0355-7; List $15; Order code HAPPENING/3HAP-5	**What's Happening in the Mathematical Sciences,** Volume 2; 1994; 51 pages; Softcover; ISBN 0-8218-8998-2; List $10; Order code HAPPENING/2HAP-5	**What's Happening in the Mathematical Sciences,** Volume 1; 1993; 47 pages; Softcover; ISBN 0-8218-8999-0; List $10; Order code HAPPENING/1HAP-5

For many more publications of interest,
visit the AMS Bookstore:

www.amsbookstore.org

AMS BOOKSTORE

Contact the AMS at 1-800-321-4AMS (4267), in the U.S. and Canada, or 1-401-455-4000; fax: 1-401-455-4046; email: cust-serv@ams.org; American Mathematical Society, 201 Charles Street, Providence, RI 02904-2294